WORLD METEOROLOGICAL ORGANIZATION

TECHNICAL NOTE No. 165

THE PLANETARY BOUNDARY LAYER

by

G. A. McBean, K. Bernhardt, S. Bodin, Z. Litynska, A. P. Van Ulden and J. C. Wyngaard
with contributions from B. R. Kerman, J. D. Reid, Z. Serbjan and J. L. Wolmsley

Edited by
G. A. McBean
Chairman, CAS Working Group on Atmospheric Boundary-layer Problems

U.D.C. 551.510.522

WMO - No. 530

Secretariat of the World Meteorological Organization - Geneva - Switzerland
1979

ISBN 92 - 63 - 10530 - 8

N O T E

The designations employed and the presentation of material in this publication do not imply the expression of any opinion whatsoever on the part of the Secretariat of the World Meteorological Organization concerning the legal status of any country, territory, city or area, or of its authorities, or concerning the delimitation of its frontiers or boundaries.

Editorial note: This publication is reproduced directly from the text submitted by the chairman of the CAS Working Group on Atmospheric Boundary-layer Problems.

CONTENTS

Page

FOREWORD

Most human activity takes place within that part of the atmosphere next to the earth's surface which we call the atmospheric boundary layer. The processes of the boundary layer are of extreme importance both for the large-scale dynamics of the atmosphere and for a large number of applications such as in meso-scale studies, evaporation, urban climatology, agrometeorology and air pollution dispersion.

This Technical Note provides a contemporary review of the current theoretical knowledge and applications of boundary-layer meteorology, stressing the recent, important advances. It will serve as an excellent source of information for meteorologists and others who are required to deal with problems related to the numerous complexities of the boundary layer.

This work has been produced by the members of the Commission for Atmospheric Sciences' Working Group on Atmospheric Boundary-Layer Problems consisting of G.A. McBean (Canada), Chairman; K. Bernhardt (German Democratic Republic); S. Bodin (Sweden); Z. Litynska (Poland); A.P. Van Ulden (Netherlands) and J.C. Wyngaard (U.S.A.). Valuable contributions were also received from B.R. Kerman (Canada); J.D. Reid (Canada); Z. Sorbjan (Poland) and J.L. Walmsley (Canada).

It gives me great pleasure to express my appreciation on behalf of the World Meteorological Organization to these scientists for their excellent work and, in particular, to Dr. McBean for editing the manuscript.

D.A. Davies

D.A. Davies
Secretary-General

SUMMARY

The planetary boundary layer is that portion of the atmosphere directly affected by the earth's surface. Within this layer (typically 1 km in depth) most human activity takes place and the processes of the boundary layer are of extreme importance both for the large-scale atmospheric dynamics and for a large number of meteorological applications. The objective of this Technical Note is to bring up-to-date technical knowledge on the boundary layer to general meteorologists.

In Chapter 1 an introduction to the general properties of the PBL is given. The status and structural aspects are reviewed with an emphasis on presenting the physical insights to problems. The Ekman layer is discussed in detail in Chapter 2. The emphasis of the Chapter is on the semi-empirical and resistance law descriptions of the Ekman layer. The topic of Chapter 3 is the tropical boundary layer. The presentation is aimed at stressing the important differences between the tropical and higher latitude boundary layers.

Chapter 4 deals with the modelling of the atmospheric boundary layer. Both research and operational types of models are reviewed. The objectives of Chapter 5 are to both introduce the reader to the observational methods used in boundary-layer meteorology and to describe briefly the major observational programmes that have been conducted.

The applications of boundary layer knowledge is the subject of Chapter 6. Three topics, the estimation of surface fluxes, aeronautical meteorology and urban boundary layers are presented in some detail while two further topics, electro-magnetic and acoustic wave propagation and problems of mountain meteorology and climatology related to the boundary layer,are more briefly reviewed.

Résumé

La couche limite planétaire est la partie de l'atmosphère qui subit directement l'influence de la surface terrestre. La plupart des activités de l'homme ont lieu dans cette couche (qui en général a 1 km d'épaisseur) et les processus qui s'y déroulent ont une extrême importance tant pour la dynamique atmosphérique à grande échelle que pour un grand nombre d'applications météorologiques. La présente Note technique a pour but de faire le point des connaissances techniques concernant la couche limite, à l'intention des météorologistes généralistes

Le premier chapitre constitue une introduction aux propriétés générales de la couche limite planétaire. L'état et la structure de la couche y sont analysés, l'accent étant mis sur les aspects physiques des problèmes. Le deuxième chapitre étudie en détail la couche d'Ekman, et traite plus particulièrement de l'approche semi-empirique et des lois de résistance qui permettent de décrire cette couche.

Le troisième chapitre est consacré à la couche limite tropicale; il est conçu de manière à mettre en évidence les différences considérables entre la couche limite des régions tropicales et celle des latitudes plus élevées.

Le quatrième chapitre a trait à la modélisation de la couche limite atmosphérique. Les deux types de modèles sont analysés: les modèles destinés à la recherche et les modèles opérationnels. Le cinquième chapitre présente au lecteur les méthodes d'observation qui sont utilisées dans le domaine de la météorologie de la couche limite et décrit brièvement les grands programmes d'observation qui ont été exécutés.

L'application des connaissances relatives à la couche limite constitue le thème du chapitre 6. Trois sujets (évaluation des flux au voisinage du sol, météorologie aéronautique et couches limites urbaines) sont traités en détail, tandis que les deux autres (propagation des ondes électromagnétiques et sonores - problèmes de météorologie et de climatologie en montagne liés à la couche limite) sont examinés plus brièvement.

РЕЗЮМЕ

Планетарный пограничный слой является той частью атмосферы, которая находится под непосредственным влиянием поверхности земли. В рамках этого слоя (типично на глубину 1 км) проходит большая часть деятельности человека, и процессы пограничного слоя имеют крайне важное значение как для крупномасштабной атмосферной динамики, так и для большого количества метеорологических применений. Цель этой технической записки довести современные технические знания о пограничном слое до широкого круга метеорологов.

В главе 1 содержится вводная часть, в которой описываются общие свойства пограничного планетарного слоя. Делается обзор аспектов состояния и структуры с упором на представление проблем с физической точки зрения. В главе 2 подробно излагается слой Экмана. В главе дается описание полуэмпирических законов и законов сопротивления слоя Экмана. В главе 3 рассматривается вопрос о тропическом пограничном слое. В главе делается попытка подчеркнуть важное различие между тропическими и высокоширотными пограничными слоями.

В главе 4 рассматривается вопрос о меделировании атмосферного пограничного слоя. Дается обзор как научно-исследовательских, так и оперативных типов моделей. Цель главы 5 заключается в том, чтобы ознакомить читателя с методами наблюдений, используемыми в метеорологии пограничного слоя, или кратко описать важные программы наблюдений, которые были проведены.

Применение знаний о пограничном слое является предметом обсуждения главы 6. Подробно представляется три вопроса; расчет приземных потоков, авиационная метеорология и городские пограничные слои и проводится краткий обзор двух других вопросов, распространение электромагнитной и акустической волны и проблемы горной метеорологии и климатологии, связанные с пограничным слоем.

RESUMEN

Constituye la capa límite planetaria aquella parte de la atmósfera afectada directamente por la superficie de la tierra. La mayor parte de las actividades humanas y de los procesos que tienen lugar en el interior de esa capa (de 1 km de espesor aproximadamente) revisten una importancia extrema tanto para la dinámica de la atmósfera a gran escala como para gran número de aplicaciones meteorológicas. La finalidad de la presente Nota Técnica no es otra que facilitar a los meteorólogos en general información técnica actualizada sobre la capa límite.

El Capítulo 1 se consagra a la presentación de las propiedades de la capa límite planetaria. En el mismo se examinan las características y aspectos estructurales de esa capa, exponiéndose en particular la naturaleza de los problemas que se plantean a ese respecto. La capa de Ekman es objeto de un estudio detallado en el Capítulo 2, en el que se exponen en particular las leyes semiempíricas y de resistencia que rigen en la capa de Ekman. El tema del Capítulo 3 lo constituye la capa límite tropical. La presentación de este tema tiene por objeto poner de manifiesto las diferencias importantes que existen entre la capa tropical y las capas límites de las latitudes más septentrionales.

El Capítulo 4 versa sobre la cuestión de la elaboración de modelos de simulación de la capa límite atmosférica, y en el mismo se examinan los modelos para fines de investigación, así como los modelos de tipo operacional. La finalidad que se persigue en el Capítulo 5 es presentar al lector los métodos de observación que se emplean en la meteorología de la capa límite y exponer sucintamente los principales programas de observación llevados a cabo.

La aplicación de los conocimientos de la capa límite constituye el tema del Capítulo 6. Tres temas - la evaluación de los flujos de superficie, la meteorología aeronáutica y las capas límites urbanas - se exponen con algún detalle, mientras que dos otros temas - la propagación de ondas electromagnéticas y acústicas y los problemas de la meteorología y la climatología de las regiones montañosas y su relación con la capa límite - se examinan de manera más sucinta.

PREFACE

The planetary boundary layer has been studied extensively during this century but is not yet completely understood. However, strong advances have been made in recent years in areas ranging from understanding of its underlying physics to our ability to observe its structure. In this Technical Note it is intended to give a contemporary, rather than historical view of the atmospheric boundary layer, stressing these recent advances. In doing so, the excursions into detail will be minimized in order to maintain a broad perspective. Therefore, at many points references are included so that the reader can pursue a more detailed study.

Because of the wide range of topics pertaining to the planetary boundary layer and the rapid advances in the science there is not a truly comprehensive up-to-date text on the PBL. There are some texts, however, that do provide much valuable information, such as Monin and Yaglom (1971); Laikhtman (1970); Zilitinkevich (1970); Lumley and Panofsky (1964); Tennekes and Lumley (1972) and the Workshop on Micrometeorology volume edited by Haugen (1973). There is, of course, extensive literature on the subject in journals such as: Journal of the Atmospheric Sciences, Boundary-Layer Meteorology, Quarterly Journal of the Royal Meteorological Society, Izvestia Atmospheric and Oceanic Physics and Beiträge zur Physik der Atmosphäre.

In spite of the tremendous advances in our understanding of the PBL in the last few decades, it seems our understanding is now accelerating. The tropical marine boundary layer is one example. It has been studied extensively recently in the GARP Atlantic Tropical Experiment, GATE. A GATE Workshop in July 1977 at the National Centre for Atmospheric Research in Boulder, Colorado, drew together over 150 participants for three weeks of discussion on tropical boundary-layer structure and processes. Other GARP experiments such as JASIN and AMTEX have or will also add greatly to our knowledge and understanding.

Many of the recent advances in our understanding are due to our increased ability to observe the atmospheric boundary layer, particularly with remote sensors. The Wave Propagation Laboratory of the U.S. National Oceanic and Atmospheric Administration has recently erected a 300 m tower (see following page) in order to develop and test remote sensors as well as to study the structure of the PBL. This facility, which is open to visiting scientists, should stimulate even greater progress toward fuller understanding of the PBL.

The planetary boundary layer is and will continue to be an exciting and challenging area for atmospheric research. In the following Chapters of this Technical Note on the Planetary Boundary Layer we hope some of this excitement will come through and that the reader is stimulated to further study and the utilization of some of the many facets of the PBL.

The 300-m tower at the Boulder Atmospheric Observatory (BOA). The tower, with eight levels of fine-resolution turbulence instrumentation, a large, fast digital data acquisition and processing system, and an array of remote sensors has been operational since early in 1978.

References

Haugen, D.A., (ed.), 1973: Workshop on Micrometeorology, American Meteor. Soc., Boston, 392 pp.

Laikhtman, D.L., 1970: Physics nf the Atmospheric Boundary Layer. Gidrometizdat Press, Leningrad.

Lumley, J.L. and H.A. Panofsky, 1964: The Structure of Atmospheric Turbulence. Interscience-Wiley, New York, 239 pp.

Monin, A.S. and A.M. Yaglom, 1971: Statistical Fluid Mechanics: Mechanics of Turbulence, Vol. 1. The MIT Press, Cambridge, Ma., 769 pp.

Tennekes, H. and J.L. Lumley, 1972: A First Course in Turbulence. The MIT Press, Cambridge, Ma., 300 pp.

Zilitinkevich, S.S., 1970: Dynamics of the Atmospheric Boundary Layer. Gridrometizdat Press, Leningrad.

CHAPTER 1

INTRODUCTION

1.1 General Properties of the Planetary Boundary Layer

The planetary boundary layer (PBL), as the name suggests, is that portion of the atmosphere directly affected by the earth's surface. The effects are traceable to the laws of heat, momentum, and mass transfer, which require continuity in properties at boundaries. When warm air flows over a cooler surface, for example, the air in contact with the surface must have the surface temperature. Above the surface the air temperature increases with height, gradually blending into the free-air temperature. Thus the vertical distribution of temperature has a boundary layer, a region where the temperature is adjusting to the surface temperature. Similarly the laws of momentum transfer require zero air velocity immediately above a stationary surface. In our simple example above, the velocity profile will also show a boundary layer.

In the case of laminar fluid motion over a flat surface of different temperature, one can calculate exactly the distributions of velocity and temperature within the boundary layer (Schlichting, 1960). The atmospheric boundary layer, on the other hand, has properties which so complicate the governing equations that solutions are much more difficult.

One of the most conspicuous and important properties of the PBL is that its flow is turbulent, not laminar. The velocity, temperature, humidity, and other properties in a turbulent flow are random functions of space and time, as anyone who has seen a detailed anemometer record, for example, will recognize. As a result, one uses a statistical approach to the calculation of PBL structure.

The turbulent nature of the PBL is perhaps the largest obstacle to its understanding and prediction. The equations governing statistical flow properties have important differences from the laminar flow equations, and cannot be solved exactly. Good approximate solutions are very expensive computationally, and in practice one resorts to model calculations; that is, one uses approximate models of the exact governing equations. Unfortunately, it is very difficult to assess the accuracy of these model calculations which is one reason why there are so many PBL models.

Thus, as with any turbulent flow, our understanding of the PBL comes in large part from observations. Model calculations provide insight, but models must constantly be tested against observations and revised when proven deficient.

Observations indicate that the structure of PBL turbulence is strongly influenced by surface conditions, temperature in particular. If the surface is warmer than the PBL air, turbulent convective circulations are generated. The regular circulation cells familiar in liquids heated over the kitchen stove, for example, are unstable to small disturbances when they occur in the PBL, and break down into random, convectively-driven turbulence.

Should the surface and PBL air temperatures be the same, there are no buoyancy effects and hence no convection; this is called the neutrally-stratified case. Historically, it was the first studied, and in spite of the absence of convection, this neutral PBL is also turbulent. This is because shear flow (that is, flow with a velocity gradient in the vertical), at PBL scales and speeds cannot remain laminar; the infinitesimal disturbances which inevitably occur are unstable, growing in space and time, forcing the flow into a turbulent state.

Should the surface be cooler than the overlying PBL air (as over land at night), a statically stable density gradient is established. We call this the "stable" PBL. It too is turbulent, but it is apt to be thinner and have less intense turbulence than its daytime counterpart.

Flow above the PBL can also be turbulent, but only intermittently in time and space; breaking gravity waves, or shearing instabilities can cause this "clear air turbulence". Thus, another definition of the PBL is the lower atmospheric region which is continuously turbulent.

Turbulence is a key feature of the PBL. Turbulent flows have far more diffusive ability than laminar ones; we recognize this in our everyday lives by stirring cream into our coffee, rather than allowing it to mix purely by molecular diffusion. Turbulence diffuses momentum and heat, as well as material; while pollutant dispersal might be the most familiar turbulent diffusion process in the boundary layer, the vertical diffusion of momentum, heat, and water vapour plays an essential role in the coupling between the earth's surface and global atmospheric circulations. We will discuss next how the structure of turbulence, the agent responsible for this vertical diffusion, differs in the characteristic states of the PBL.

1.2 States of the PBL

The PBL is almost constantly evolving in response to both the diurnal heating cycle and to changing synoptic conditions. As a result its structure and depth can vary considerably over space and time. However, it does typically have distinct states that we can idealize somewhat and discuss in fairly simple terms. Our idealizations are primarily the assumption of a horizontal, uniform surface and the neglect of strong time changes, such as those caused by frontal passages.

Our usual classification of PBL states is based on buoyancy effects. If a parcel of air is adiabatically displaced upward (in the z direction), it will expand because of the reduced pressure aloft and hence its temperature, T, will decrease. The resulting temperature profile satisfies, for dry air,

$$\frac{\partial T}{\partial z} = -\frac{g}{C_p} \qquad (1.2.1)$$

where C_p is specific heat at constant pressure. This is called the adiabatic temperature profile. The PBL is then neutrally buoyant; displaced parcels have exactly the density of their surroundings, and thus experience no net buoyant force. We call this neutral stratification. Should the mean temperature decrease with height more slowly than the adiabatic profile, a vertically displaced parcel will experience a force tending to restore it to its original position. We call this stable stratification. The final case where the temperature decrease with height exceeds the adiabatic lapse rate is called unstable stratification; here displaced parcels are vertically accelerated away from their original positions.

In fact it is the density of a displaced parcel that we should be considering, and as such the water vapor distribution also enters; but it is traditional to deal with the temperature profile and generalize, as will be discussed later, to virtual temperature. However, these profiles actually indicate static stability, and none is in practice a completely reliable indicator of the true state of the PBL. It is found, for example, that in the mid regions of the convective PBL the temperature profile often indicates slightly stable stratification. We can usually rely on surface layer profiles as indicators of the PBL state, however, since they usually reflect the surface conditions.

1.2.1 The Convective PBL

During clear days a land surface is normally warmer than the air aloft, because of heating by incoming solar radiation. This warmer, near-surface air is buoyant and establishes convective, turbulent motions. In some situations, over water for instance, near-surface air becomes buoyant because it contains more water vapor (which is less dense than air) than does the air at upper levels. These two effects can be treated mathematically in a unified way by treating the air-water vapor mixture as a perfect gas. Its pressure P, gas constant R, density ρ and "virtual" temperature T_V satisfy

$$P = \rho RT_V \quad (1.2.2)$$

(Lumley and Panofsky, 1964). The gas constant is that for dry air, and T_V is related to true temperature T through the specific humidity q (mass of water vapor per mass of moist air)

$$T_V = T (1 + 0.61 q) \quad (1.2.3)$$

We see that density changes at constant pressure can be caused either by actual temperature changes or by changes in the specific humidity; in other words, buoyant air is either warmer or more humid, or both, than its surroundings.

Boundary-layer convection would extend to much greater heights than observed were there not usually an overlying layer of stably stratified air at a level z_i which typically ranges from a few hundred m to a few km. This "inversion" layer acts as a lid for the convection by damping vertical motions, and establishes the depth of the convective PBL.

This inversion lid can be eroded from below by turbulence and displaced vertically by a mean velocity, such as that induced by convergence or divergence in the horizontal wind field. Therefore, the convective PBL normally becomes deeper as the day progresses, because of turbulent entrainment of air down into the PBL; however in some instances its depth can be held stationary or even lowered by subsidence. The latter situation can cause air pollution episodes by trapping pollutants in an abnormally thin PBL.

At times the inversion lid is penetrated by convection, gravity waves, or other events, thus releasing the PBL content to the free atmosphere above. This can create deep convection extending to the tropopause (~ 20 km) and is often associated with deep cumulus formation and severe thunderstorms.

While the roots of PBL convection are near the surface, the turbulent motion it drives is most intense in the mid-regions of the PBL. At lower levels the boundary inhibits motions while higher up the damping influence of the lid is felt. Although the fluctuating wind, or eddy, structure is random it can contain recognizably discreet elements such as plumes, roll-like horizontal vortices, and vertically oriented vortex motions such as dust devils.

A fluid in motion at velocity $\underset{\sim}{u}$ responds to a variety of forces, including those due to buoyancy. Spatial velocity gradients create inertia forces proportional to $(\underset{\sim}{u}.\underset{\sim}{\nabla})\underset{\sim}{u}$, according to the Navier Stokes equations (Tennekes and Lumley, 1972). These equations also tell us that a fluctuation in either the vertical pressure gradient or in the buoyancy force per unit mass, $\frac{g}{T_o}\,\theta$, will produce a local vertical acceleration. Here θ is the local deviation from the mean temperature T_O. This assumes the Boussinesq approximation (Lumley and Panofsky, 1964), which is valid for the low Mach number motion on the PBL scale. Consider now a roughly circular eddy motion in a vertical plane in the PBL, extending over a scale $\ell \sim 10^3$m with an eddy velocity $q \sim 1\ \mathrm{ms}^{-1}$. Inertia forces will be of order q^2/ℓ; if buoyancy forces are of the same order, we have

$$\frac{q^2}{\ell} \sim \frac{g}{T_o}\,\theta \qquad (1.2.4)$$

Using our values, we find $\theta \simeq 0.03$ C. This is small, and apt to be found routinely; thus we expect that buoyancy effects to be important relative to inertial effects in the convective PBL.

It should be no surprise, then, that a very small surface temperature excess is observed to drive the PBL into convection. The resulting intense vertical mixing flattens out the vertical profiles of mean wind, temperature, and humidity above the surface layer. At $z \sim z_i$ the profiles often show sharp jumps to free-atmosphere values. Typical examples are sketched in Figure 1.1. Following the usual convention, we have plotted potential temperature Θ, defined by

$$\frac{\partial\Theta}{\partial z} = \frac{\Theta}{T}\left(\frac{\partial T}{\partial z} + \frac{g}{C_p}\right) \qquad (1.2.5)$$

Note that Θ is independent of z in the neutral case, from Eq. (1.2.1). To account for moisture it is usual to define a virtual potential temperature, $\Theta_v \equiv \left(\frac{\Theta}{T}\right) T_v$. Broadly speaking, Θ_v in the PBL plays the same role as ordinary temperature T in a small-scale (laboratory) boundary layer.

Note also in Figure 1.1 the three distinct regions in the vertical. The near-ground, or surface layer has strong gradients in wind, temperature, and moisture. This layer is typically a few tens of meters deep, tending to be thinner

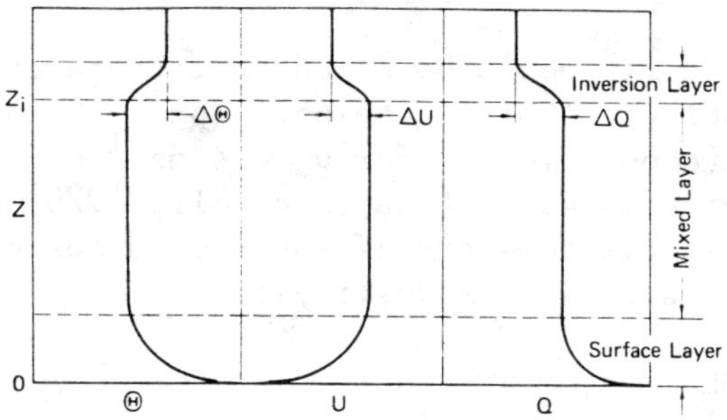

Figure 1.1 A schematic of the mean vertical profiles of potential temperature, wind speed, and specific humidity in the convective PBL. Note the strong gradients in the surface and inversion layers and the flat profiles in the mixed layer.

but more intense the stronger the convection. Above this, and extending nearly to the inversion base at z_i, is the mixed layer. As the name suggests, the mean gradients are usually smallest here. Above this is the inversion layer where the turbulence is extinguished by the stable stratification and the profiles approach the conditions of the free atmosphere. Figure 1.2 is an artist's conception of the structure of the convective PBL, showing the eddy motions and the capping inversion above.

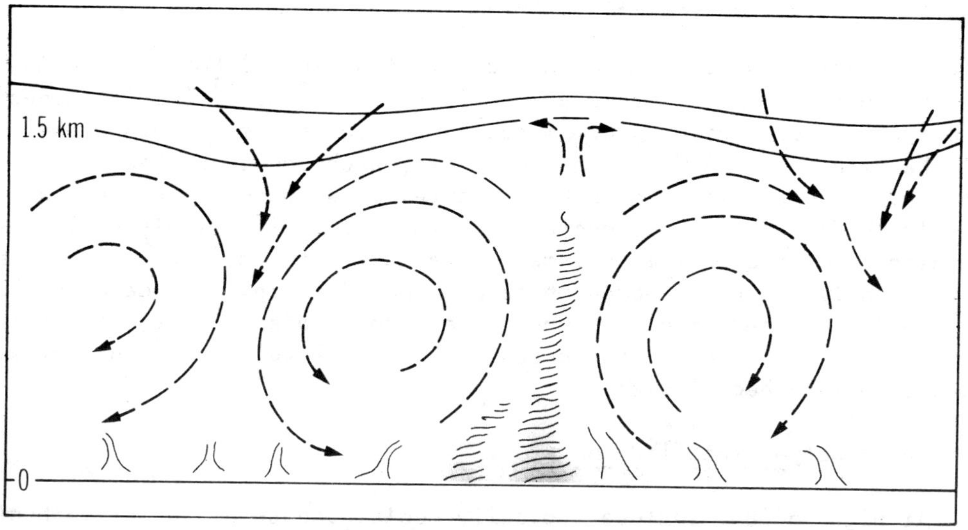

Figure 1.2 An artist's conception of the structure of the convective PBL. Note the thin capping inversion layer at 1.5 km height, and the entrainment of this layer by the large eddies below. Near the surface are convective plumes, some reaching near the top of the layer.

For completeness it should be noted that there exists beneath the surface layer, a viscous sublayer that is laminar, not turbulent. This layer is very thin, of order 5 ν/u_* (about 0.01 mm for $u_* \sim .4$ ms^{-1} and air). The viscous sublayer can exist over near calm water (Hupfer et al., 1975) or other very smooth surfaces. When the roughness elements are of the same order or more than the depth of the viscous sublayer, the sublayer is destroyed.

1.2.2 The Neutral PBL

If the PBL has an adiabatic lapse rate throughout (Θ = constant, which can happen if the surface moisture and heat fluxes are negligible and there is no inversion aloft) we have the neutral case. Here turbulence is due entirely to the wind shear (the change in wind velocity with height) and there are no buoyancy effects. In view of our previous calculation showing that only small temperature fluctuations give strong bouvancy effects in the PBL, one must wonder about the likelihood of occurrence of a truly neutral case. Nevertheless, it has been widely studied theoretically.

According to theoretical studies, the neutral PBL establishes its depth h in proportion to u_*/f. The friction velocity u_* is equal to $\sqrt{\tau_o/\rho}$ with τ_o the surface stress and ρ the air density; f is $2\Omega \sin \phi$, with Ω the earth's rotation rate and ϕ the latitude. One sometimes sees statements that $h = 0.25\ u_*/f$, although the constant will depend somewhat on how one judges the depth. In contrast to the convective case, here the PBL properties blend gradually into the free atmosphere, and turbulence levels gradually fall to zero.

Theoretical studies also indicate that a neutral PBL is affected by an inversion at heights $z_i \simeq u_*/f$ or lower. Using $f = 10^{-4}$ sec^{-1} (midlatitudes), and taking a typical value of $u_* = 0.4$ m sec^{-1}, we see that an inversion below 4 km will affect the structure. A downward heat flux (which can be caused by a cool surface) can have strong effects on PBL structure, as we discuss in the next section. It is difficult to determine experimentally if the PBL is in a truly neutral state unless sophisticated turbulence measurements are made, and this was not possible until fairly recently. Thus although there are many "neutral" PBL data in the literature, one cannot be certain to what extent they refer to the true neutral case. Some data are undoubtedly from cases with upper level inversions, or small but significant surface heat fluxes.

1.2.3 The Stably-Stratified PBL

At night a land surface typically cools because of radiative heat loss to space. The near-surface air cools and creates a positive (stable) temperature gradient in the PBL. This has strong dynamical effects on the turbulence and hence on the structure of the layer. Energy must be expended to maintain vertical velocity fluctuations in the presence of the stable lapse rate; since turbulence is inherently three-dimensional, with energy exchanges taking place among all three velocity components, the effect of extraction of energy from the vertical motions is transmitted to the horizontal components as well.

We can look at this energy extraction in stably-stratified flow with the model we used earlier. Imagine a circular eddy motion of scale ℓ in a vertical plan. Its motion will induce temperature deviations of order $\ell\cdot(\partial\Theta_v/\partial z)$, and the eddy will feel a vertical restoring force (per unit mass) of order $\frac{g}{\Theta_v}\frac{\partial\Theta_v}{\partial z}\ell$. The rate of working against this restoring force (i.e., the rate of extraction of eddy kinetic energy) is of order $\frac{g}{\Theta_v}\frac{\partial\Theta_v}{\partial z}\ell q$. Eddy inertia forces are of order q^3/ℓ, and thus the ratio of energy extraction (through working against buoyancy) to energy input (through inertia effects) is

$$\frac{\frac{g}{\Theta_v}\frac{\partial\Theta_v}{\partial z}\ell^2}{q^2} \qquad (1.2.6)$$

This means that, all things being equal, the largest eddies will be damped the most strongly; sufficiently small scale eddy motions will not be directly affected by a stable lapse rate. However, since all scales in turbulence are dynamically coupled, because of the nonlinearity of the Navier-Stokes equations, a stable lapse rate will affect all scales of motion.

Equation (1.2.6) defines an "eddy Richardson number" - ratio of buoyancy and inertia forces on a given size eddy. A more familiar variety is the gradient Richardson number

$$Ri = \frac{\frac{g}{\Theta_v}\frac{\partial\Theta_v}{\partial z}}{\left(\frac{\partial V}{\partial z}\right)^2} \qquad (1.2.7)$$

which can be interpreted as a Richardson number for the dominant, energy-containing turbulent eddies. Ri is often used as a stability indicator since it is relatively easy to measure. Note Ri is positive in the stable case, negative in the unstable case, and zero under neutral conditions.

Observations do indicate that the scale of turbulence in the stably-stratified PBL is smaller than in the neutral or convective cases (Kaimal et al., 1972). They also clearly indicate that stable PBL's tend to be thinner, and support smaller fluxes of momentum, heat, and moisture. Figure 1.3 illustrates some of the salient features of the stably stratified PBL.

1.3 Structural Aspects

We mentioned that an important feature of the PBL (or any boundary layer) is the diffusion of momentum, heat, and mass across it. Molecular action alone causes this diffusion in a laminar flow. For example, in a laboratory boundary layer having gradients only in the vertical (z) direction, the vertical fluxes in the laminar case are:

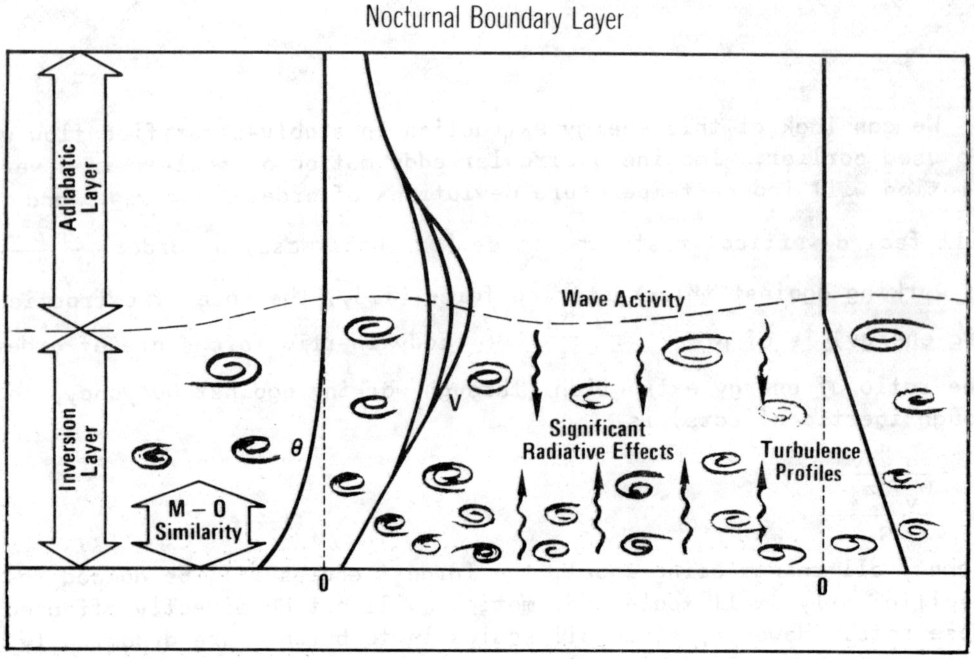

Figure 1.3 Some of the salient features of a nocturnal boundary layer. Its upper edge is characterised by wave activity and an inertial oscillation (nocturnal jet) in the wind profile. The turbulent eddies in the layer are relatively small; the structure near the surface follows Monin-Obukhov similarity.

$$\tau = -\nu \frac{\partial U}{\partial z} \rho$$

$$H = -\kappa \frac{\partial T}{\partial z} \rho \; C_p$$

$$M = -D \frac{\partial Q}{\partial z} \rho \qquad (1.3.1)$$

where ν, κ, and D are the molecular diffusivities (units of length squared per unit of time) of momentum, temperature, and mass (water vapor, for example); U, T, and Q are the horizontal speed, temperature, and water vapor concentration; and ρ and C_p are the density and specific heat at constant pressure. In this case of molecular diffusion, the diffusivities are properties of the fluid.

In the turbulent PBL, the fluxes are largely due to the turbulence itself. To illustrate this, consider a simple case where the velocity, temperature, and water vapor concentration vary monotonically and smoothly with z. Since the flow is turbulent, the velocity at a point will vary randomly about its mean value U (z); likewise, the virtual potential temperature and humidity will vary randomly about Θ_v (z) and Q (z). Let u, w, θ_v, and q represent these fluctuations in horizontal and vertical velocity, virtual potential temperature, and water vapor. Note, if Θ/T is significantly different from unity then a correction factor T/Θ must be multiplied to Θ_v. If the fluctuations u, θ_v, and q are correlated with the vertical fluctuation w, there will be turbulent fluxes. The total fluxes will be:

$$\tau = \left(-\nu \frac{\partial U}{\partial z} + \overline{uw}\right) \rho$$

$$H_v = \left(-\kappa \frac{\partial \Theta_v}{\partial z} + \overline{w\theta_v}\right) \rho \, C_p$$

$$M = \left(-D \frac{\partial Q}{\partial z} + \overline{wq}\right) \rho \qquad (1.3.2)$$

where the overbar represents an average.

We can crudely relate these turbulent fluxes to the mean gradients. Taking the virtual temperature flux as an example, in the presence of a mean virtual potential temperature gradient a particle displaced vertically a distance d by a vertical velocity fluctuation w will generate a temperature fluctuation $\theta_v \sim -\partial \Theta_v/\partial z \cdot d$. Thus,

$$\overline{w\theta_v} \simeq -\overline{wd} \, \frac{\partial \Theta_v}{\partial z} \qquad (1.3.3)$$

The covariance $\overline{wd}$ should be positive, since positive (negative) d is caused by positive (negative) w. Hence, $\overline{wd}$ is evidently an "eddy diffusivity", K, of order $q\ell$, where q and ℓ are velocity and length scales of the energy-containing eddies. Thus, extending these arguments to moisture and momentum transfer, we have:

$$\tau \simeq -(\nu + K) \frac{\partial U}{\partial z} \rho$$

$$H \simeq -(\kappa + K) \frac{\partial \Theta}{\partial z} \rho \, C_p$$

$$M \simeq -(D + K) \frac{\partial Q}{\partial z} \rho \qquad (1.3.4)$$

Note that K is property of the flow, not of the fluid. Note also that the ratio of turbulent and molecular diffusivities, in the case of momentum, is

$$\frac{K}{\nu} \simeq \frac{q\ell}{\nu} = Re \qquad (1.3.5)$$

which is a turbulent Reynolds number. For $q = 1 ms^{-1}$, ℓ = 100m, and air, we have

$$\frac{K}{\nu} \simeq \frac{q\ell}{\nu} \simeq 7 \times 10^6 \qquad (1.3.6)$$

Since for air $\nu \simeq \kappa \simeq D$, we see that turbulent diffusion in the PBL is far more important than molecular diffusion.

Turbulent transfer processes are much more complicated than our simple model indicates, and relations such as Eq. (1.3.4) are valid only in the simplest flows (Tennekes and Lumley, 1972). Buoyancy effects, for example, can invalidate these simple arguments. Nonetheless, it is true that turbulent fluxes are far larger than molecular fluxes in the PBL, except immediately at the surface in the viscous sublayer, and hereafter by <u>fluxes</u> we will mean the turbulent fluxes.

Turbulent fluxes have important roles in the equations for mean values of horizontal velocity, temperature, or humidity. In the horizontally uniform case, (with only mean pressure varying with x and y) these equations are

$$\frac{\partial U}{\partial t} + \frac{\partial}{\partial z}\overline{uw} = -\frac{1}{\rho_o}\frac{\partial P}{\partial x} + fV \qquad (1.3.7)$$

$$\frac{\partial V}{\partial t} + \frac{\partial}{\partial z}\overline{vw} = -\frac{1}{\rho_o}\frac{\partial P}{\partial y} - fU \qquad (1.3.8)$$

$$\frac{\partial \Theta}{\partial t} + \frac{\partial}{\partial z}\overline{\theta w} = 0 \qquad (1.3.9)$$

$$\frac{\partial Q}{\partial t} + \frac{\partial}{\partial z}\overline{qw} = 0 \qquad (1.3.10)$$

where P, the mean pressure field, can be taken as imposed by the large-scale synoptic conditions above the PBL. We have neglected phase changes and radiative effects. The second term in each equation is turbulent flux divergence; this term cannot in general be neglected. Equation (1.3.9), for example, says that the mean temperature at a point changes in time only because of a divergence in the turbulent heat flux; in other words, it changes only because of turbulent mixing. The turbulent terms must be known in order to solve for the distributions, in time and space, of mean wind, temperature and humidity. This is sometimes called the fundamental problem in turbulence: prediction of the turbulent fluxes so mean fields can be calculated.

The pressure gradient terms in (1.3.7) and (1.3.8) are often written as:

$$f\,V_g = \frac{1}{\rho_o}\frac{\partial P}{\partial x} \qquad (1.3.11)$$

$$f\,U_g = -\frac{1}{\rho_o}\frac{\partial P}{\partial y} \qquad (1.3.12)$$

where U_g and V_g are the geostrophic wind components. Above the boundary layer the steady flow satisfies $U = U_g$, $V = V_g$, since the turbulent momentum flux divergence is negligible there. Within the PBL the strong friction insures that $U \neq U_g$, $V \neq V_g$; as a result there is a wind spiral in the PBL, with both speed and direction changing between the surface and the free atmosphere. This is usually referred to as the Ekman spiral. In the Northern (Southern) Hemisphere, looking down from above, the wind rotates clockwise (counter-clockwise) with increasing height. This spiral can be strongly modified by stability effects, horizontal inhomogeneity, height-dependent pressure gradients, or time changes.

We can derive some simple integral constraints on PBL wind profiles. Starting from steady, horizontally homogeneous conditions, where (1.3.7) and (1.3.8) reduce to

$$\frac{\partial}{\partial z}\overline{uw} = f\left(V - V_g\right) \qquad (1.3.13)$$

$$\frac{\partial}{\partial z}\overline{vw} = f\left(U_g - U\right) \qquad (1.3.14)$$

we integrate between z = 0 and z = h. We choose the surface wind to be along the x direction; hence as $z \to 0$, $\overline{uw} \to -u_*^2$ and $\overline{vw} \to 0$ (since there is no preferred lateral direction the correlation between v and w will be zero), and since both stresses vanish at z = h, we have

$$u_*^2 = f \int_0^h (V - V_g)\, dz \qquad (1.3.15)$$

$$0 = \int_0^h (U_g - U)\, dz \qquad (1.3.16)$$

Since in our simple model $(U, V) \to (U_g, V_g)$ at the PBL top, it follows from 1.3.16) that U must exceed U_g somewhere in the PBL. Equation (1.3.15) is sometimes used to determine surface stress from wind profile and geostrophic wind measurements.

Having discussed these general structural features, we can now consider some details of regions within the PBL.

1.3.1 The Surface Layer

The surface layer is the name traditionally used for the lowest few tens of meters of the PBL. It has been extensively studied in recent decades, and a comprehensive summary of that research is presented in the Workshop on Micrometeorology (Haugen, 1973).

Many of the idealizations we make for the PBL as a whole are more realistic in the surface layer. Principal among these are horizontal homogeneity and stationarity.

The horizontally homogeneous surface layer has mean properties that are independent of horizontal position; they depend only on z and time. This requires an infinite uniform surface; although these do not exist, strictly speaking, they are approached over the oceans and over broad flat prairies. The latter might be called locally homogeneous surfaces. Similarly, the surface layer can never be truly stationary, because of the inevitably changing boundary conditions. One is apt to find quasi-steady conditions, however, in which the time changes are so small as to be dynamically negligible, and the structure can be considered to be passing through a sequence of steady states.

Some simple hypotheses have proven very useful in describing and analyzing the structure of the locally homogeneous, quasi-steady surface layer. Perhaps the best-known is the Monin-Obukhov (M-O) similarity hypothesis. The surface layer is essentially a onstant flux layer (that is, at heights small compared with the boundary layer depth the turbulent fluxes are nearly equal to the surface flux values) and so there is reason to believe its structure is determined by only a few parameters. By the M-O hypothesis these key parameters are z, g/Θ_v, the kinematic surface stress τ_o/ρ and the surface virtual temperature flux $Q_{ov} = H_{ov}/\rho\, C$ (where H_{ov} is the surface virtual heat flux).

One might ask why other parameters, such as kinematic viscosity ν, Coriolis parameter f, or surface roughness length z_o are not included in the governing group. Except very near the surface, or when the smallest eddies are being considered, the viscous forces are far smaller than inertia forces because the turbulent Reynolds number is so large (Tennekes and Lumley, 1972); hence the value of viscosity is important. Coriolis forces are also small compared with inertia forces in the surface layer, so the value of f is also unimportant there. Far enough from the surface ($z >> z_o$), z_o is no longer directly important; its main effect is to determine the stress level, or drag for a given mean wind speed. Thus M-O similarity holds for $z_o << z << h$.

The M-O parameters define scales

$u_* = \sqrt{\tau_o/\rho}$ (velocity scale)

$T_* = -Q_{o_v}/u_*$ (temperature scale)[1]

$$L = -\frac{u_*^3\, \theta_v}{kg\, Q_{o_v}} \qquad (1.3.17)$$

z (length scales)

By dimensional analysis, flow properties when non-dimensionalized with u_*, T_*, and z are universal functions of z/L. For example, the mean wind shear $\partial U/\partial z$ in any quasi-steady, locally homogeneous surface layer is

$$\frac{\partial U}{\partial z} = \frac{u_*}{z} F_1\,(z/L) \qquad (1.3.18)$$

where F_1 is a universal function. The <u>functions</u> are not given by M-O theory; they must be determined by other theoretical means or by experiment. Surface-layer research data are now routinely presented in the M-O framework.

In our simple example of (1.3.18), we note that under neutral conditions $L \to \infty$, so we have

$$\frac{\partial U}{\partial z} = \frac{u_*}{z} f_1\,(0) = \frac{u_*}{kz} \qquad (1.3.19)$$

where k is the von Karman constant. This gives the well-known logarithmic wind profile

$$U = \frac{u_*}{k} \ln \frac{z}{z_o} \qquad (1.3.20)$$

with U = 0 at $z = z_o$. The logarithmic wind profile does not necessarily apply near z_o, however, as discussed above; the meaning of (1.3.20) is merely that the behaviour at $z >> z_o$, when extrapolated near the surface, gives U = 0 at $z = z_o$.

[1] Monin and Obukhov originally defined $T_* = -Q_{o_v}/ku_*$. The von Karman's constant (the value of which is in dispute) is omitted here and in many recent publications.

Note also that (1.3.19), and thus the logarithmic profile, holds only under conditions such that $|z/L| \ll 1$.

One can also write M-O predictions analogous to (1.3.18) for the vertical gradients of temperature and moisture. Yaglom (1977) has recently reviewed the present information concerning the shape of the universal functions (see also Section 6.1). Viewed another way, these expressions indicate that the surface layer eddy diffusivity is $k u_* z \, F_2 \, (z/L)$, where the stability function F_2 can be different for momentum or scalar (temperature, moisture) transfer. Note for momentum that $F_2 \, (0) = 1.0$.

In general both the temperature and moisture fluxes contribute to buoyancy. Thus we use the surface virtual heat flux

$$Q_{o_v} = Q_o + 0.61 \; T \; M_o \qquad (1.3.21)$$

instead of Q_o as a M-O governing parameter. This distinction is not normally critical over land surfaces, but is usually important over water, where $0.61 \; T \; M_o$ can be much larger than Q_o. In fact, in flow over a very slightly cooler water surface, Q_o is negative (down) while M_o is positive; thus the PBL could be convective, driven by the moisture flux. The surface layer temperature profile, meanwhile, would seem to indicate a stable PBL. Clearly here one must deal with virtual temperature.

The M-O similarity structure takes simpler forms in the limiting cases of very unstable (z/L large negative) and very stable (z/L large positive) conditions. These cases are discussed in detail in the Workshop volume.

Not all surface layer statistics follow M-O similarity. The mean vertical gradients of wind, temperature, and moisture, and the statistics of vertical velocity and temperature fluctuations seem to, but the horizontal velocity statistics do not. Evidently the horizontal velocity fluctuations in the surface layer can be influenced by large eddies that extend to the top of the PBL. Thus they are not always determined simply by the variables in the M-O governing group. This is discussed to some extent in the Workshop volume and also in a recent paper by Panofsky et al. (1977). Evidence for deviations from M-O similarity is now easier to obtain than before, because of advances in our boundary-layer measurement capability, and the next few years should see much clarification here.

Real surface layers will depart to some extent from the idealizations inherent in M-O theory. There are several reasons including:

1) time changes in boundary conditions
2) inhomogeneity
3) radiative effects
4) non-flat terrain

Taking the first of these, we note that the mean horizontal pressure gradient and the vertical heat flux are almost always changing in time. Turbulent flows can respond to such changes with a time scale $\tau \simeq \ell/q$. In the surface layer $\ell \simeq z$, and $q \simeq u_*$; taking $z = 30$ m, $u_* = 0.3 \; ms^{-1}$, we find $\tau \simeq 100$ sec. Boundary condition changes will usually have a much longer time scale than this, and so the surface-layer turbulence should be able to follow these changes perfectly. Thus we can speak of "quasi-steady" surface layer turbulence, meaning that it goes through essentially steady states as it changes in time.

The inhomogeneity question is more difficult. If surface-layer air encounters different surface conditions downstream (such as different roughness or temperature) it is forced to adjust its structure. Far enough downstream after such a step change the surface layer will again be in equilibrium with the near surface stress and heat flux. Clearly the streamwise distance required for this response is determined by the internal dynamics of the turbulent flow. Therefore this response distance will depend on the state of the surface layer (neutral, stable, unstable) and the nature of the change, since all these will influence the turbulence structure and thus its ability to respond to change. From many model calculations, and some experiments, we can say that neutral surface layer structure makes an essentially complete response to a roughness change in a horizontal distance about 100 times the height above ground. This distance should be less in convective conditions and greater in stably-stratified conditions. Little is known about the details of response to changing surface conditions in any but this simplest case, and this clearly is an area that needs more attention, both theoretically and experimentally.

The air within the surface layer can be cooled or heated not only through turbulent heat flux divergence, but also through the divergence of radiative flux. This is not accounted for in the M-O hypothesis. It is difficult to determine how important an influence radiative flux divergence has on the surface-layer structure. It is known, however, that at night over land the minimum temperature is often not at the surface, but slightly above; simple radiative models (Funk, 1960) do predict strong cooling at heights of the order of 20 cm. Gavrilov and Laikhtman (1973) have examined the interaction of the vertical profiles and the long wave radiative fluxes. Some work has been done on the influence on the turbulence structure (Coantic and Seguin, 1971). The radiative properties of air depend on humidity as well as temperature, so the problem is quite complicated and also deserves more study.

Perhaps the most difficult question to answer relates to the extent to which surface layer structure is influenced by hilly terrain. There have been some experimental studies (see Weber, 1975, for example) and some model calculations (see Gent and Taylor, 1976, for example), but this subject is still relatively unexplored.

1.3.2 Upper Levels

The structure above the surface layer depends strongly on the stability. A convenient stability index here is h/L, where h is the PBL depth and L is the M-O length. A neutral PBL has h/L = 0; if h/L is negative, the PBL is unstable or convective; if h/L is positive, the PBL is stable. We will now briefly survey upper-level structure in these three cases.

1.3.2.1 Convective Conditions

The indications (primarily from numerical model calculations) are that very small -h/L values (perhaps near 1) are sufficient to drive the PBL into a convective state. In the limiting state of free convection the wind and the surface stress vanish. It is usual to define convective mixed layer velocity and temperature scales as

$$w_* = \left(\frac{g}{\theta_o} Q_{ov}\ h\right)^{1/3} \qquad (1.3.22)$$

$$\theta_* = \left(\frac{Q_{ov}^{\ 2}\ \theta_o}{gh}\right)^{1/3} \qquad (1.3.23)$$

and the indications are, from model calculations (Deardorff, 1972) and field measurements (Kaimal et al., 1976) that turbulence statistics aloft do scale with w_* and θ_* under convective conditions even in the presence of mean wind. Recall from our discussion in Section 1.2 that the turbulence is driven by buoyancy, and its intertial forces develop in response so that $q^2/\ell \sim g\theta/T$. Taking $q \sim w_*$, $\ell \sim h$, $\theta \sim \theta_*$ we see that, from (1.3.22) and (1.3.23)

$$\frac{q^2\ T_o}{\ell\ g\ \theta} \sim \frac{w_*^{\ 2}\ T_o}{h\ g\ \theta_*} = 1 \qquad (1.3.24)$$

which confirms that inertia and buoyancy forces are of the same order.

Velocity fluctuations in the convective PBL can be quite large. At mid-latitudes in summer, over land, we might have $Q_{ov} = 0.3$ ms^{-1}C, h = 2 km, which would give (1.3.22) $w_* \sim 2.6$ ms^{-1}. Energetic eddies fill the PBL, extending to the surface (where as mentioned earlier they disturb M-O similarity for horizontal wind fluctuation statistics) and to the inversion base at z_i, where they erode it by entraining air from above. The eddy diffusivity K in the convective PBL can be expected to be order $z_i w_*$, which can be much larger than in the neutral case. In the example just given, $z_i w_* \sim 5 \times 10^3$ m s^{-1}. This greatly enhanced vertical mixing gives flat mean wind and temperature profiles above the surface (as shown in Figure 1.1); this is often called a "mixed layer".

The stress profiles in an idealized convective mixed layer can be explained quite simply. When the mean momentum (1.3.7) and (1.3.8) are differentiated with respect to z, we find in steady state

$$\frac{\partial^2}{\partial z^2}\ \overline{uw} = -f\ \frac{\partial V_g}{\partial z} \qquad (1.3.25)$$

$$\frac{\partial^2}{\partial z^2}\ \overline{vw} = f\ \frac{\partial U_g}{\partial z} \qquad (1.3.26)$$

since the wind shear can be neglected. Therefore in a barotropic (geostrophic wind independent of z) PBL, the stress curvature vanished and the stress profiles are linear. Then if we choose coordinates with the x-direction aligned with the surface wind, we have, if the stresses vanish at z = h,

$$\overline{uw} = u_*^2 \ (1 - z/h) \qquad (1.3.27)$$

$$\overline{vw} = 0 \qquad (1.3.28)$$

The stress profiles predicted by (1.3.27) and (1.3.28) are not often seen for three reasons. First, fixed-point measurements (such as those from towers or balloon cables) require very long averaging times to give statistically reliable stresses (Wyngaard, 1973); several hours or more may be required, and the PBL is not often stationary over that length of time. Thus fixed-point data usually have much scatter. A second reason is that the PBL often has a significant amount of baroclinity, or geostrophic wind shear. Horizontal temperature gradients cause the horizontal pressure gradient to vary with z according to the thermal wind equation:

$$f \frac{\partial V_g}{\partial z} = \frac{g}{T_o} \frac{\partial T}{\partial x} \qquad (1.3.29)$$

$$f \frac{\partial U_g}{\partial z} = - \frac{g}{T_o} \frac{\partial T}{\partial y} \qquad (1.3.30)$$

A surprisingly small horizontal temperature gradient will cause significant geostrophic wind shear and stress profile curvature. Taking $\partial^2\overline{uw}/\partial z^2 \simeq \Delta\overline{uw}/h^2 \simeq 0.5\ u_*^2/h^2$ (that is, a $\overline{uw}$ profile curved so its departure from linearity at mid-levels is $0.5\ u_*^2$) gives, from (1.3.25) and (1.3.29)

$$\left|\frac{\partial T}{\partial x}\right| = \frac{T_o}{g} \frac{\partial^2 \overline{uw}}{\partial z^2} \simeq \frac{1}{2} \frac{T_o}{g} \frac{u_*^2}{h^2} \qquad (1.3.31)$$

The use of $T_o \simeq 300$ K, $u_* \simeq 0.3$ ms $^{-1}$, $h \simeq 10^3$m gives, from (1.3.31),

$$\left|\frac{\partial T}{\partial x}\right| \simeq \frac{0.15 \text{ C}}{100 \text{ km}} \qquad (1.3.32)$$

This is extremely small; much larger gradients are common. Thus we can expect truly barotropic PBL's to be about as rare as truly neutral ones.

A final reason for the departure of stress profiles from the linear forms of (1.3.27) and (1.3.28) is a velocity jump across the inversion at z_i. Deardorff (1973) has shown that velocity jumps ΔU and ΔV will induce stress jumps $\Delta\overline{uw}$ and $\Delta\overline{vw}$ according to

$$\Delta\overline{uw} = -w_e \, \Delta U \qquad (1.3.33)$$

$$\Delta\overline{vw} = -w_e \, \Delta V \qquad (1.3.34)$$

where w_e is the entrainment velocity at the interface; that is, the rate at which the inversion base is being eroded or its height changed,

$$w_e = \frac{dh}{dt} - w_h \qquad (1.3.35)$$

where w_h is the mean vertical velocity at z = h. If w_e = 3 cm s^{-1}, (100 m hr^{-1}) which is typical of growing convective PBL's, and ΔU = 3 ms $^{-1}$, then $\Delta\overline{uw}$ = 0.1 m^2 s^{-2}, which is as large as typical surface stresses (u_* = 0.3 ms^{-1}). Then the profile will be <u>linear</u> in the PBL between $-u_*^2$ at the surface and $\Delta\overline{uw}$ at z_i, and fall sharply to zero in the inversion.

If we can neglect the effects of horizontal temperature gradients and velocity jumps at z = h we can extract some useful information from our mixed layer model. In coordinates aligned with the surface layer wind, we have above the surface layer but within the mixed layer

$$V \simeq 0$$

$$U \simeq U_g \qquad (1.3.36)$$

Since $V \simeq 0$ and the $\overline{uw}$ profile is linear (1.3.27) in the mixed layer, the U equation of motion (1.3.7) and (1.3.15) require that

$$V_g \simeq \frac{u_*^2}{fh} \qquad (1.3.37)$$

Hence within the capping inversion there will be a small amount of turning as V goes from zero in the mixed layer to $-u_*^2/fh$ above. For typical values of u_* = 0.3 ms^{-1}, f = 10^{-4} s^{-1}, h = 1 km, U_g = 10 ms^{-1}, we have

$$V_g \simeq -\frac{u_*^2}{fh} \simeq 1\ \mathrm{ms}^{-1}$$

$$\alpha = \tan^{-1}\frac{V_g}{U_g} \simeq 6\ \mathrm{deg.} \qquad (1.3.38)$$

In this simple model of the convective PBL most of the wind shear is concentrated in the surface layer. There the wind shear is M-O similar (see 1.3.18), and there are empirical expressions for it (see Workshop volume). If we assume the surface layer thickness scales with L, we can integrate the shear expression across the surface layer to find the wind speed at the outer edge. In our case this is U_g, and since we showed $V_g \ll U_g$, this speed is essentially $G = (U_g^2 + V_g^2)^{1/2}$. Wyngaard et al. (1974) in this way found the convective layer drag law

$$\frac{G}{u_*} = \frac{1}{k}\left[\ell n - \left(\frac{L}{z_o}\right)\right] \qquad (1.3.39)$$

which agrees fairly well with Wangara observations (Clarke et al., 1971).

The well-mixed nature of the convective PBL has some simple implications for the vertical distributions of mean values and vertical fluxes of humidity and temperature. Figure 1.1 shows idealized profiles of mean humidity and potential temperature. Note the strong gradients in the surface layer, the flat profiles in the mixed layer and the jumps $\Delta\theta$ and ΔQ at the inversion base at $z = z_i = h$. Figure 1.4 shows some data from the AMTEX experiment (described in a paper by Wyngaard et al., 1977) that show similar behaviour. The humidity jump can be of either sign but the one shown is more common. Lilly (1968) has shown that these jumps, in conjunction with entrainment of the inversion by the PBL convection, generate jumps in the fluxes at z = h. If we denote the flux values just below the inversion base by a subscript h, his results imply

$$\overline{w\theta})_h \simeq -w_e\,\Delta\Theta \qquad (1.3.40)$$

$$\overline{wq})_h \simeq -w_e\,\Delta Q \qquad (1.3.41)$$

where w_e is the entrainment velocity introduced earlier.

Within the mixed layer the intense turbulent activity drives the mean gradients of potential temperature and humidity to quasi-steady state. Thus by differentiating these mean equations we find, under horizontally homogeneous conditions

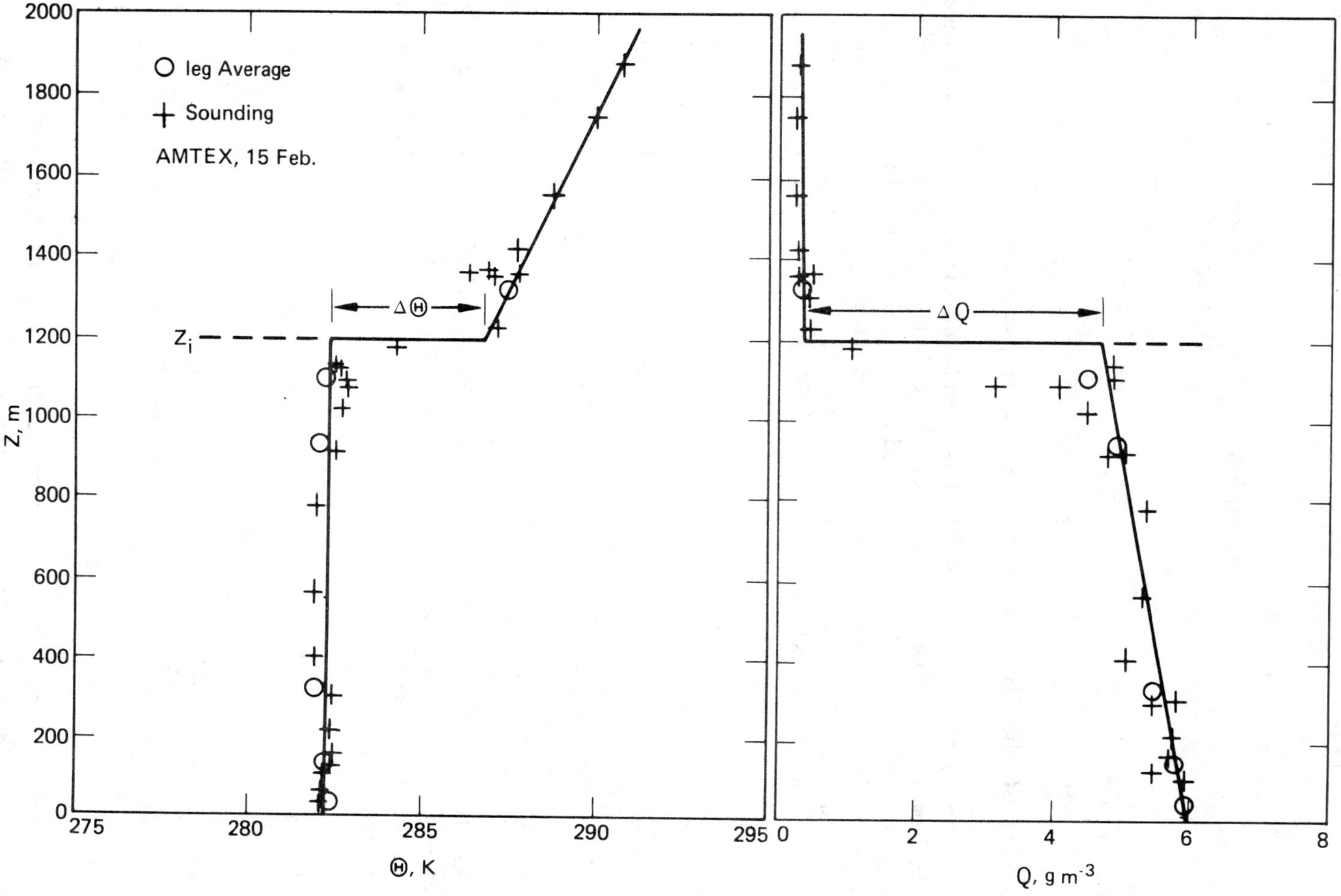

Figure 1.4 Vertical profiles of mean potential temperature and mean absolute humidity as measured over the sea in the AMTEX experiments. Note the strong jumps in both properties at z_i.

$$\frac{\partial}{\partial t}\frac{\partial \Theta}{\partial z} = -\frac{\partial^2}{\partial z^2}\overline{\theta w} \simeq 0 \qquad (1.3.42)$$

$$\frac{\partial}{\partial t}\frac{\partial Q}{\partial z} = -\frac{\partial^2}{\partial z^2}\overline{qw} \simeq 0 \qquad (1.3.43)$$

giving linear profiles of the fluxes of moisture and temperature.

In some cases the convective PBL has scattered clouds or a cloud layer near its top. This complicates the PBL structure by bringing in phase change thermodynamics (the energy associated with the vapour-liquid water transition is large) as well as radiative heat flux divergence (clouds have drastically different radiation properties from clear air). We cannot pursue the details of this important case here but the reader is referred to the studies of Lilly (1968) and Deardorff (1976) for more information.

1.3.2.2 Neutral Conditions

We mentioned in Section 1.2 that the PBL can be driven into convective or stably-stratified states by small surface - PBL temperature differences. We also mentioned that an inversion below a few km height can also strongly affect PBL structure. Both conditions are common, and so the neutral PBL is apt to be of more academic than practical interest.

An important feature of the neutral PBL is the shift in wind direction across it. We can exmaine this by considering first the equations of motion for the flow above the PBL, which we assume to be non-turbulent and horizontally homogeneous (except for mean pressure):

$$\frac{\partial U}{\partial t} = f\left(V-V_g\right) = fV - \frac{1}{\rho_o}\frac{\partial P}{\partial x} \qquad (1.3.44)$$

$$\frac{\partial V}{\partial t} = f\left(U_g-U\right) = -fU - \frac{1}{\rho_o}\frac{\partial P}{\partial y} \qquad (1.3.45)$$

Figure 1.5 shows an assumed steady-state mean pressure field for the Northern Hemisphere (if positive). Note that $\partial P/\partial x$ is positive; $\partial P/\partial y$ is zero, so from (1.3.45) U must be zero in steady state. Hence the steady-state free, or geostrophic, flow above the PBL is parallel to the isobars, as shown in Figure 1.5 and has magnitude $G = (U_g^2 + V_g^2)^{1/2}$. In the Southern Hemisphere (f negative) the direction differs by 180°.

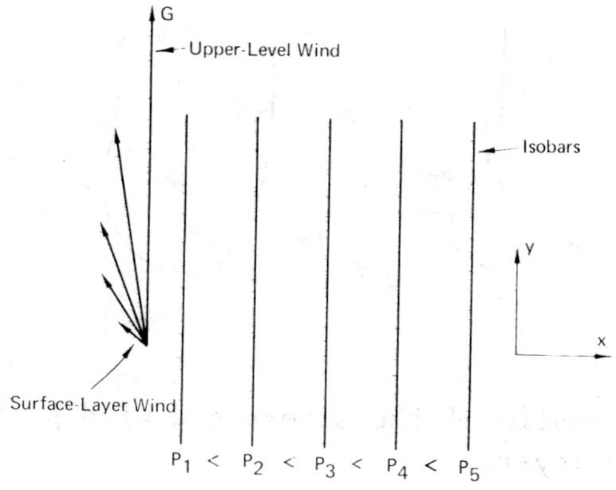

Figure 1.5 A schematic of the relation between the mean pressure field and the mean wind in an idealized Ekman layer. Note that the upper-level wind is parallel to the isobars, while the near-surface wind is at an angle in this Northern Hemisphere example.

Within the PBL, turbulent friction causes the wind direction shift with respect to the isobars. The PBL wind satisfies (in the coordinate system in Figure 1.5):

$$\frac{\partial U}{\partial t} + \frac{\partial}{\partial z}\overline{uw} = f\left(V - V_g\right) \qquad (1.3.46)$$

$$\frac{\partial V}{\partial t} + \frac{\partial}{\partial z}\overline{vw} = f\left(U_g - U\right) \qquad (1.3.47)$$

Consider the V-equation (1.3.47). Friction acts to retard the PBL wind; hence $\partial\overline{vw}/\partial z$ will be positive. Since f U_g is zero, fU must be negative to balance the friction. Hence U is negative in the PBL (northern Hemisphere). Similarly, from the U-equation (1.3.46) we conclude that the friction term will be negative; if f V_g is constant, fV must be smaller than in the free atmosphere. Thus we have the wind spiral sketched in Figure 1.5, with the wind rotating clockwise (counter-clockwise), between the surface and the free atmosphere, looking down from above in the Northern (Southern) Hemisphere.

We can now deduce much about the wind and stress profiles within the neutral PBL from (1.3.46) and (1.3.47). We revert to our usual coordinate system, with x along the surface layer wind direction so that U_g is positive and V_g is negative (Northern Hemisphere). In the surface layer $V \simeq 0$, and (1.3.46) shows that $\partial\overline{uw}/\partial z$ is positive. Note that f $(V-V_g)$, and hence $\partial\overline{uw}/\partial z$, is maximum at the surface. In the surface layer $U < U_g$, so $\partial\overline{vw}/\partial z$ is also positive. If the turbulent stress in the surface layer lies along the mean wind direction then $\overline{vw} = 0$ near the surface (since there should be no preferred lateral direction there). We also have the integral constraints for the profiles (1.3.15) and (1.3.17). Thus we can sketch the stress and wind profiles in Figure 1.6. The sign reversal in $\overline{uw}$ near the PBL top is shown in model calculations and is evidently due to the corresponding reversal in sign of $\partial U/\partial z$. The model calculations (e.g. Deardorff, 1972); Wyngaard et al. 1974) also indicate turbulence distributions monotonically decreasing with z.

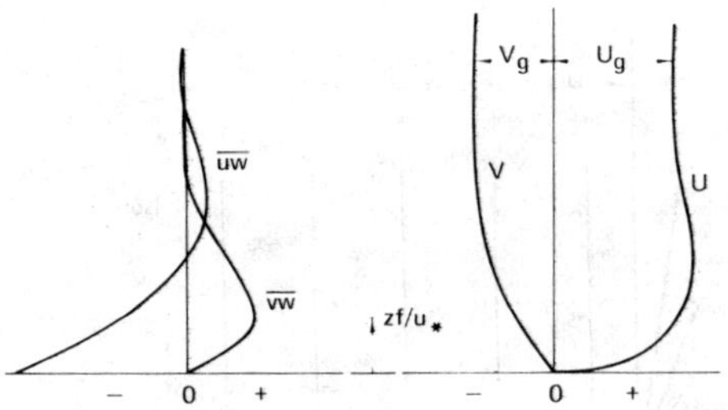

Figure 1.6 A schematic of the stress and wind profiles in an idealized Ekman layer.

Some elegant theoretical work (Kazanskii and Monin, 1961, Blackadar and Tennekes, 1968) has provided much insight into the structure of the neutral PBL. It is usual now to speak of the "geostrophic drag law":

$$\ln \frac{G}{fz_o} = B + \ln\left(\frac{G}{u_*}\right) + \left[\frac{k^2 G^2}{u_*^2} - A^2\right]^{1/2} \qquad (1.3.48)$$

where A and B are presumably universal constraints,

$$A = -k\frac{V_g}{u_*} \qquad (1.3.49)$$

$$B = \ln\left[\frac{u_*}{fz_o}\right] - k\frac{U_g}{u_*} \qquad (1.3.50)$$

but some authors reverse the definitions of A and B. Theoretical and numerical results indicate that the scale height for the neutral PBL is u_*/f. This approach is discussed more fully in Chapter 2.

There is an important class of "neutral" PBL's that have distinctly different structure. Here the PBL is capped by a low inversion at height h and the surface heat flux is negligible. The stable stratification in the inversion layer extinguishes the turbulence at a height h much less than the expected thickness ($\simeq u_*/f$) of the neutral PBL. We consider here the simplest idealization of this case, where the entrainment velocity w_e at the inversion base is negligible. Thus, from (1.3.33) and (1.3.34) the stresses at z = h vanish. We see from the equations of motion, (1.3.46) and (1.3.47), that friction is of order u_*^2/h while Coriolis forces are of order f u_*. Thus Coriolis forces are $hf/u_* \ll 1$ times inertia forces, and can be neglected; the equations of motion give immediately the stress profiles

$$\frac{\partial}{\partial z}\overline{uw} \simeq -fV_g \rightarrow \overline{uw} = -u_*^2\left(1 - \frac{z}{H}\right) \qquad (1.3.51)$$

$$\frac{\partial}{\partial z}\overline{vw} \simeq 0 \rightarrow \overline{vw} = 0 \qquad (1.3.52)$$

Note that in this limit the PBL flow is parallel to the pressure gradient, and hence at 90° to the flow above. Note also that with constant pressure gradient $u_*^2/h \simeq -fV_g$ = constant, so $u_*^2 \propto h$, and $u_*^2 \to 0$ as $h \to 0$. In this limit we have a large jump in wind speed as well as 90° direction change across the inversion at z = h.

This case has been discussed in more detail by Csanady (1974). Typical situations in nature are apt to involve some entrainment at z = h, and hence be not as simple as our example. Zeman and Tennekes (1977) discuss some aspects of the entrainment problem.

1.3.2.3 Stable Conditions

The stably-stratified PBL is quite different in structure from the neutral or convective cases. It tends to be thin; depths of a few tens to a few hundred m are commonly observed in acoustic sounder traces, while the convective PBL is usually at least an order of magnitude thicker. The surface fluxes of momentum, heat, and moisture are perhaps an order of magnitude less than in the convective case. Turbulence levels are much lower, and as we saw earlier the largest eddies are strongly damped, generally limiting the dominant eddy size ℓ. Since $K \sim q\ell$, turbulent diffusivities are also much smaller than in neutral or convective cases.

The negative vertical gradient of virtual potential temperature PBL leads to wave activity not seen in the neutral or convective cases. When $\partial\theta_v/\partial z$ is negative, a parcel of air displaced vertically experiences a buoyant force tending to return it to its original location. When pressure gradient, advective, and viscous forces are neglected the equation of motion for small displacements z is

$$\frac{d^2z}{dt^2} = -\frac{g}{T_o}\frac{\partial\theta_v}{\partial z} z \qquad (1.3.53)$$

Stable oscillations are possible for only $\partial\theta_v/\partial z$ positive, in which case they occur at the Vaisala-Brunt frequency $N = \left(\frac{g}{T_o}\frac{\partial\theta_v}{\partial z}\right)^{1/2}$. In the stable PBL the oscillation period $2\pi/N$ lies roughly in the range 30-300 sec. Acoustic sounder records from the nocturnal PBL often show striking evidence of waves with periods in this range. These gravity waves are governed by equations more complicated than our simple example, but the Vaisala-Brunt frequency N is an important parameter governing their structure and behaviour.

The stably-stratified PBL differs in two other important ways from neutral or convective PBL's. One difference is the sensitivity of its structure to sloping terrain. To illustrate this consider a uniform, flat surface sloped at an angle α to the horizontal. If we rotate our coordinate axes so that the z-direction is normal to the surface and the x and y axes are parallel to it, then the horizontal equation of motion will have gravity force terms proportional to sin α. If we adopt the usual decomposition of the mean temperature field into a background adiabatic profile T_o (z) and a deviation $\overline{T}'$ (z) (see Lumley and Panofsky, 1964) then the streamwise mean momentum equation is, in the horizontally homogeneous case,

$$\frac{\partial U}{\partial t} + \frac{\partial}{\partial z}\overline{uw} = f\left(V-V_g\right) + \frac{g}{T_o}\overline{T}' \sin\alpha \qquad (1.3.54)$$

The new term, the last on the right, represents the drainage acceleration, or the tendency for cool air to be accelerated downslope and decelerated upslope. Surprisingly small slopes can have large effects. For example, take $h = 100$ m, and $u_*{}^2 = 10^{-2}m^2\ s^{-2}$ so the friction term is of order $10^{-4}\ s^{-2}$. If the PBL air is 3°C cooler than the adiabatic profile, $\alpha \sim 10^{-3}$ (a slope of 1 part in 1000) gives a drainage acceleration as large as the turbulent friction term. This effect does not exist in the neutral PBL, since $\overline{T}' = 0$.

We see from this simple example that the stably-stratified PBL is very sensitive to sloping terrain, which in effect acts to make the driving horizontal pressure gradient height-dependent. Thus we expect a barotropic model of the stably-stratified PBL to be less useful over land than over water, and probably not typically as good as approximation as in the convective case.

There is also reason to believe that the "quasi-steady" approximation, which is often useful for descrbing the convective PBL, is less relevant to stable conditions. Recall that the convective PBL has large eddies with characteristic velocity w_* which fill the layer of depth z_i. Thus an "eddy turnover time" is of order z_i/w_*, typically 5-10 min. If the time scale of the changes in boundary conditions is a few hours, as is typical, then we expect the structure to follow these changes perfectly and remain in quasi-steady state. Data are consistent with this interpretation. A stable PBL, on the other hand, has relatively smaller, less energetic eddies which do not communicate information on changing surface conditions so quickly to the rest of the layer even though h must be 1/10 of what z_i had been. As the surface temperature budget and hence the surface heat flux changes, for example, we can estimate the time required for the layer as a whole to respond by considering an eddy diffusion time. If the layer has an average eddy diffusivity K and a thickness h, the time T for a boundary condition change to diffuse to height h is of order h^2/K. Typical values are $h \sim 100$ m, $K \sim 0.3^2 m\ s^{-1}$; this gives $T \sim 10$ hr. Recall that the convective PBL, $K \sim z_i\ w_*$ so that $T \sim z_i/w_*$, typically 5-10min. By contrast, the T value at night is so large that under many circumstances we must expect the nocturnal PBL over land to be far from quasi-steady.

Thus the non-steadiness that characterizes the early hours of development of a noctural PBL often persists through the night. Over land, the development process typically begins shortly before sunset, when the surface heat loss exceeds the gain by solar radiation. Model calculations (e.g., Delage, 1974; Wyngaard, 1975) indicate that if the cooling rate of the near-surface air is constant then a steady-state stable PBL can be established within a few hours over a horizontally uniform, flat surface. If radiative flux divergence is ignored (which model studies suggest is often valid except very near the surface) the mean potential temperature in this idealized case satisfies

$$\frac{\partial \Theta}{\partial t} + \frac{\partial}{\partial z}\overline{\theta w} = 0 \qquad (1.3.55)$$

and the calculations indicate that in the lower half or so of the layer the steady - state solution behaves as

$$\frac{\partial \Theta}{\partial t} = -\frac{\partial \overline{\theta w}}{\partial z} \simeq \frac{Q_o}{h} \qquad (1.3.56)$$

with both the surface temperature flux Q_o and the PBL depth h essentially time-independent.

A similar situation can be expected in air-mass modification with warm air flowing over a coller surface, for example a large, cool body of water. Here the diurnal changes can be negligible, and far enough from the land-sea discontinuity a layer in quasi-steady equilibrium can be established that satisfies

$$\frac{U\partial T}{\partial x} + \frac{\partial}{\partial z}\overline{\theta w} = 0 \qquad (1.3.57)$$

so that again

$$\frac{\partial}{\partial z}\overline{\theta w} \simeq -\frac{Q_o}{h} \qquad (1.3.58)$$

and Q_o and h vary slowly with streamwise distance x.

More typically, however, the stable PBL is apt to be out of equilibrium, either because of changes in time or streamwise distance. As an example of the former, we note that a constant near-surface cooling rate cannot usually be sustained long enough to drive the PBL to steady-state. The surface temperature adjusts in response to soil heat flux, air heat flux, and the radiative balance; as a result, the high cooling rates found in the early evening hours usually decrease with time. As an example of the latter, we note that a stable PBL responds slowly in the streamwise direction to changes at the lower boundary. After a change in roughness and temperature, as for example in the land-sea problem mentioned earlier, an equilibrium internal boundary layer grows, and far enough downstream it has thickened to the PBL depth. A rule-of-thumb depth of the internal boundary layer following a roughness change in the neutral case is 1% of the downstream distance. A stable case might respond an order of magnitude slower so conceivably a 100 m deep, equilibrium stable boundary layer in the land-sea problem might require 100 km of downstream distance to establish.

Thus idealized, quasi-steady models are probably less applicable to the real world stable PBL than to the convective PBL. Real cases are apt to be complicated by terrain slope, wave activity, non-steadiness, or horizontal inhomogeneity. With this caveat, then, we present a brief sketch of idealized stable PBL structure as revealed by model calculations.

Figures 1.7-1.9 show the evolution of u_*, the cross-isobaric angle α, and Q_o during the development of a nocturnal PBL in the evening hours. These are taken from the study by Wyngaard (1975). Note that within a few hours the PBL, which has a constant cooling rate near the lower surface, is quasi-steady. The surface stress $(-u_*^2)$ is much less than its value at the transition hour, and the surface heat flux is also much less than typical daytime values. Note that for strongly stable conditions (caused by high cooling rates) α is near 45 degrees, the value calculated by constant eddy-viscosity PBL models.

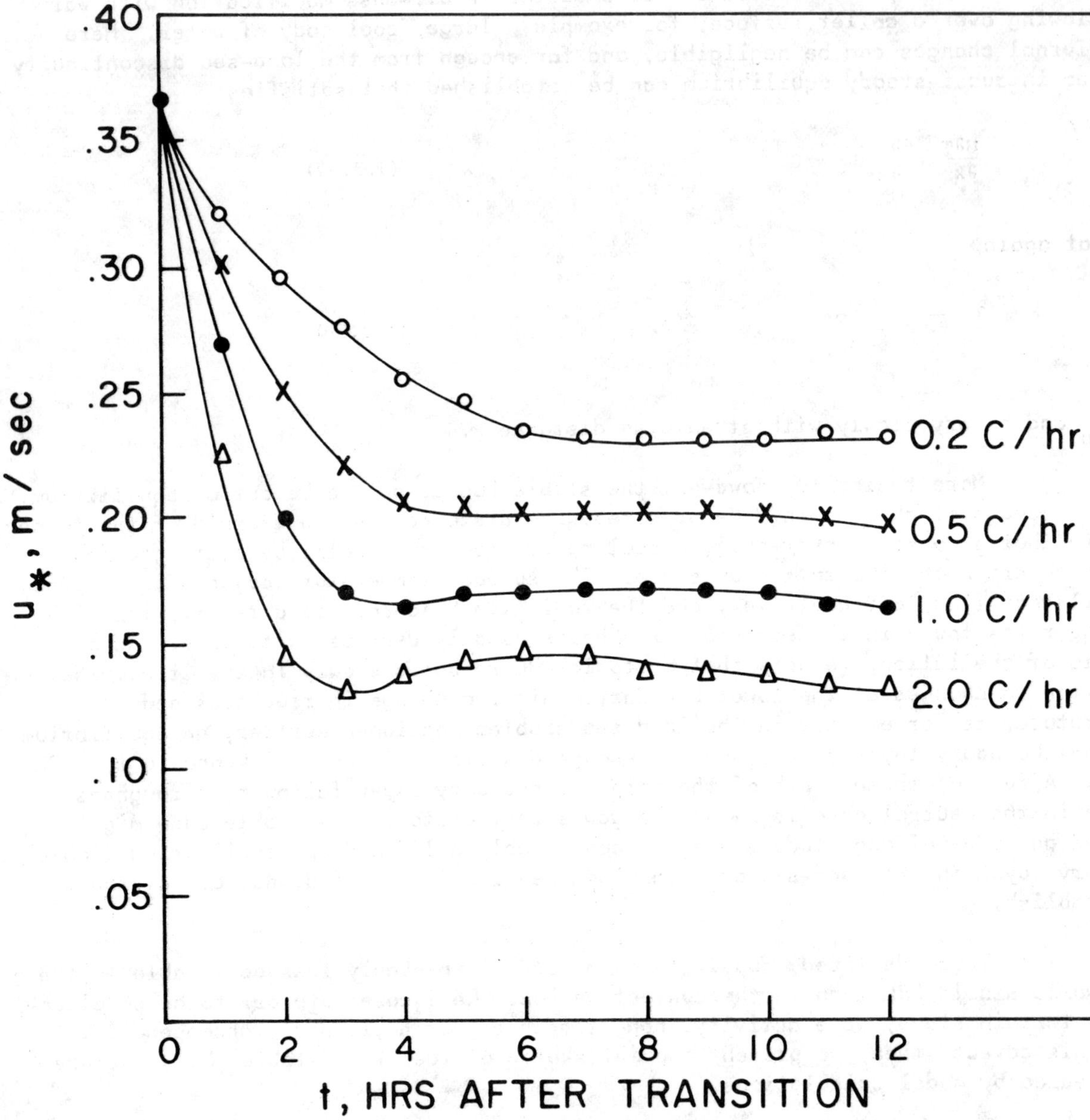

Figure 1.7 The behaviour of u_* in an evolving nocturnal PBL, as predicted by the numerical model of Wyngaard (1975). Various cooling rates of the near-surface air were used, as shown.

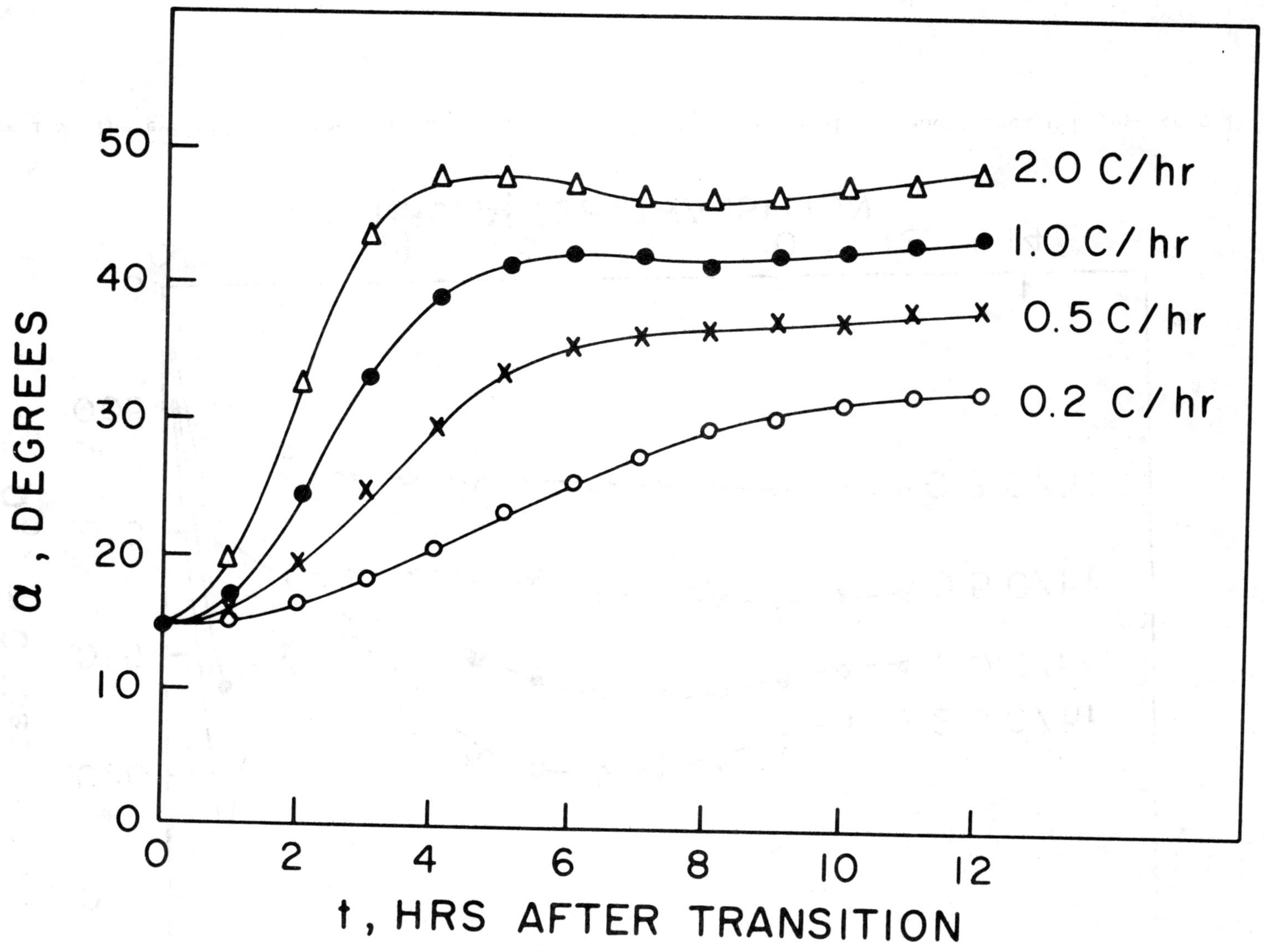

Figure 1.8 The evolution of the cross-isobaric angle α in an evolving nocturnal PBL (Wyngaard, 1975).

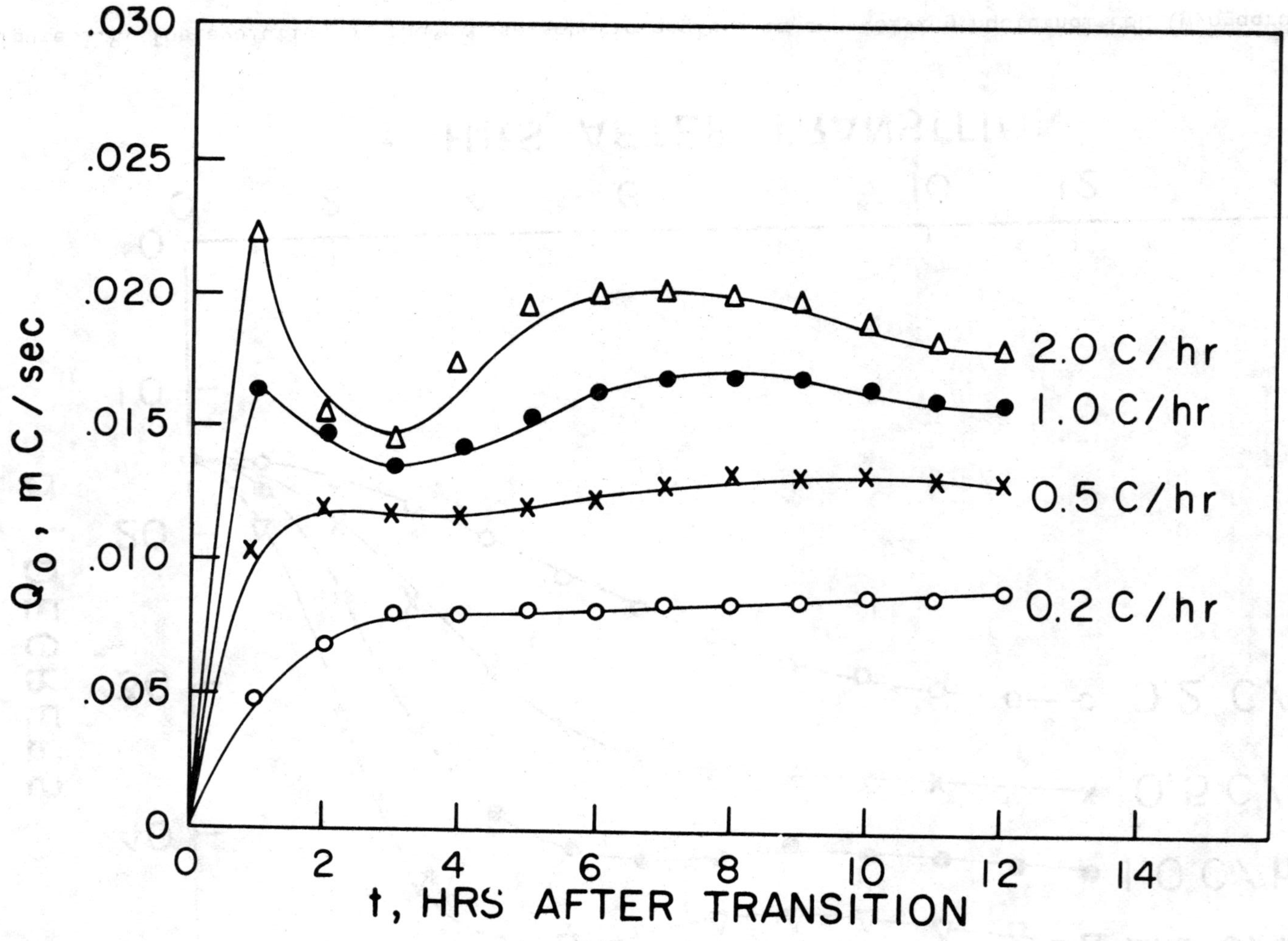

Figure 1.9 The evolution of the surface temperature flux Q_o in an evolving nocuturnal PBL (Wyngaard 1975).

Figure 1.10 shows steady-state turbulence profiles within the PBL. The depth h follows the similarity predition (Zilitinkevich, 1972)

$$h = d \left[\frac{u_* L}{f} \right]^{1/2} \qquad (1.3.59)$$

with d a constant that is rather model sensitive; Wyngaard (1975) found $d \sim 0.2$; Businger and Arya's (1974) value is about twice as large.

The top of the stable PBL is marked by a transition to non-turbulent flow, which in the real world may be complicated by intermittent, breaking wave motions. Model results indicate that both gradient and flux Richardson numbers (the ratio of buoyant extraction to shear production of turbulent kinetic energy) increase to critical values at z = h, which is associated with the vanishing of the turbulence.

1.5 Modelling and Parameterization

At the outset we should explain why, as we indicated in Section 1.1, it is necessary to model the PBL instead of solving its governing equations exactly.

The Navier Stokes equation, which expresses the momentum balance in a moving fluid (liquid or gas), is non-linear. This non-linearity links dynamically all the scales of motion (or eddies) in a turbulent flow (Batchelor, 1953). Some of the consequences of this, such as the inertial energy cascade made famous by Kolmogorov (1941) are well known. The important aspect here is that in any turbulent flow, the range of dynamically important eddy sizes extends from ℓ (of the order of the integral scale, or flow size) to η (the Kolmogorov microscale). Most of the turbulent kinetic energy is carried by ℓ-sized eddies, while the viscous dissipation occurs in the η-sized ones.

If we estimate the scale range between ℓ and η we can quickly see why a numerical solution of the Navier-Stokes equations for the PBL on a large scale, fast digital computer is impossible. In any turbulent flow ε is of the order of q^3/ℓ (Tennekes and Lumley, 1972); if we take for the PBL $q \sim 1\ ms^{-1}$, $\ell \sim 300m$, we find $\varepsilon \sim 30\ cm^2\ s^{-3}$; using ν for air, we see that $\eta = (\nu^3/\varepsilon)^{1/4} \sim 0.1$ cm. Thus a numerical grid in space must have resolution of the order of $\eta \sim 0.1$ cm and extend throughout the PBL. The grid must be three-dimensional, since turbulence is, and if $h \sim 10^3$ m we will need 10^{18} grid points to resolve the flow in a 10^3 m cube! No computer can handle such a task.

Clearly, then, straightforward numerical simulations of the PBL are out of the question. One can solve the equations with a coarser grid if the effects of the missing small eddy motions can be modelled in some way. These "sub-grid scale" eddies are responsible for the viscous dissipation, as we mentioned earlier. A model without at least a parameterization of "sub-grid scale" eddies would soon "run away" -- the turbulent kinetic energy would grow without bound.

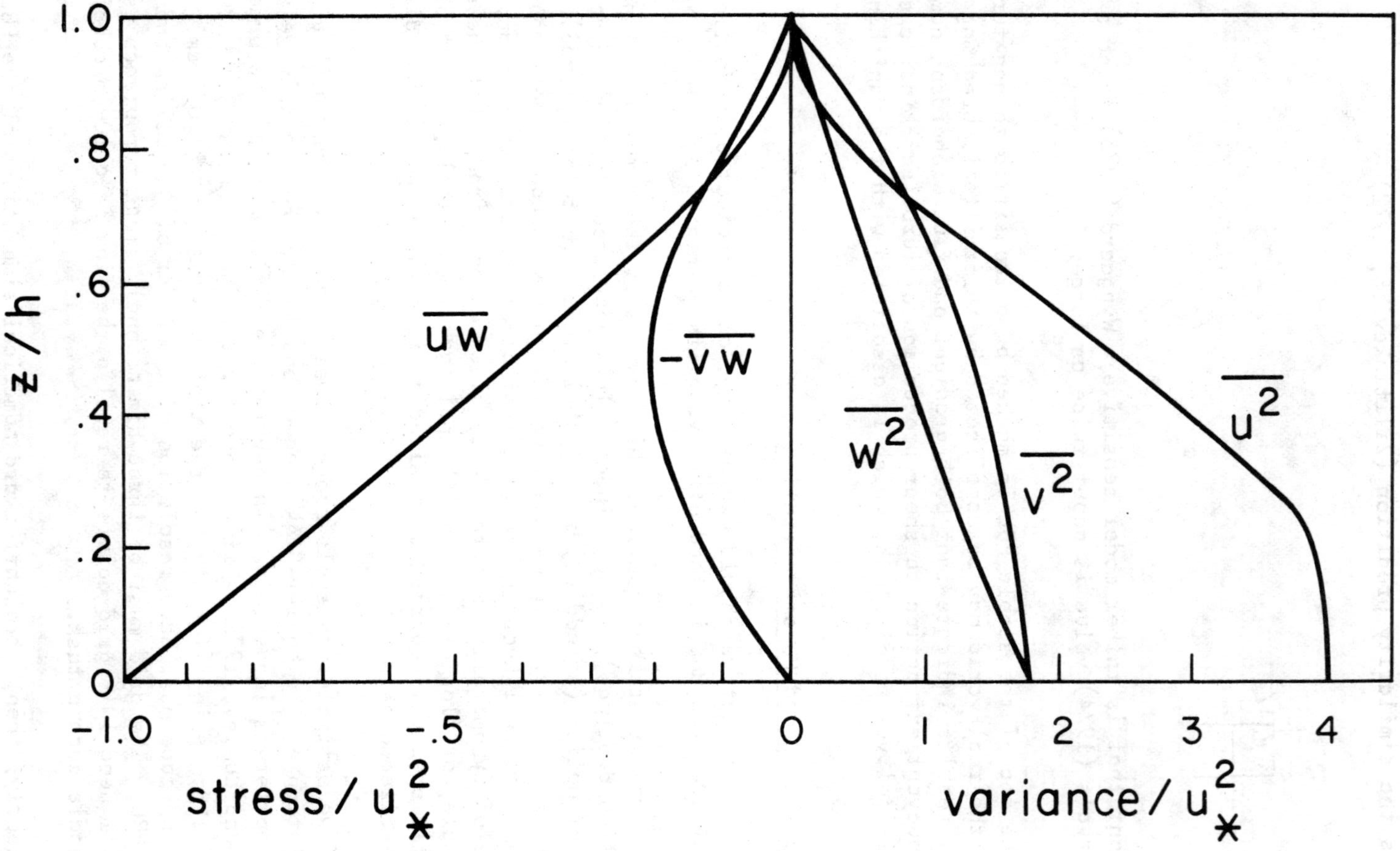

Figure 1.10 The steady-state turbulence profiles in the nocturnal PBL, as found from the model calculations of Wyngaard (1975)(Northern Hemisphere).

Deardorff has performed an extensive series of calculations with a 3-D model of the PBL. He modelled the flow in a volume using of the order of 10^4 grid points. His 1974 paper, for example, gives results from a 5 km x 2 km volume using $(40)^3$ grid points; the grid intervals were 125 m in the horizontal and 50 m in the vertical. This allows the explicit calculation of most of the energy-containing range of turbulence, but is unable to resolve any of the dissipation structure, which must therefore be parameterized. This approach is very expensive computationally; the model typically runs as much as an order of magnitude slower than real time, so a 24 hour simulation can require a few hundred hours computer time. Nevertheless, it is very successful and remains perhaps the best way, short of direct measurement, to explore the structure and dynamics of the PBL.

The 3-D approach actually calculates the fluctuating motion in the resolvable range. This gives much detail which often is not needed; often one wants only average properties, such as the turbulent fluxes or the mean profiles. An alternative is to attempt to solve averaged (over horizontal planes, or over an ensemble) equations. However, as we saw earlier, the averaging process generates new terms in the equations; the mean momentum equations contain Reynolds stress terms, for example. These are unknowns, and must be approximated (parameterized) before the equations can be solved.

Historically, the first attempts to model the PBL dealt with the averaged equations. Until recently the eddy diffusivity approximation discussed earlier was universally used in such calculations. One simply chose a K profile and solved (at first analytically, later numerically on computers) for the mean wind, temperature, or concentration profile or other quantities of interest.

A problem with this approach is the proper specification of K. Unfortunately there is no systematic method for doing this; the literature contains a myriad of K profiles that have been used. The recent work by Wipperman (1973) contains a large set of examples of K-calculations of PBL structure (see Chapters 2 and 4 for further details).

A modern alternative to K-calculations, one which is still in a stage of rapid evolution, is the so-called second-moment method. Here the set of equations for Reynolds stress, heat flux, and other second moments is "closed" by approximating third moments and other unknown terms. A discussion of this approach can be found in the Workshop volume.

A model calculation can be used for a variety of purposes, ranging from obtaining information on PBL structure for direct application to some practical problem, to studying the basic physics of the flow. One important application, which spans this range, is the development of parameterizations.

By parameterizations we mean relationships among flow variables. All models use them; K models use eddy diffusivity parameterizations, 3-D models use sub-grid scale parameterizations. Second moment models use third moment parameterizations -- that is, expressions that relate third moments to other known (or calculated) flow properties.

The systematic generation of parameterizations remains, in a sense, the fundamental problem in turbulence. In a narrower sense, however, one strong rationale for generating better and better models of turbulence, and of the PBL, is to be able to generate better parameterizations of the flow for use in simpler models for practical applications.

For example, an environmental engineer calculating diffusion in the PBL needs a simple yet effective model, and research with sophisticated models can give the parameterizations needed in such everyday models. Similarly, mesoscale models, however sophisticated, must use coarse grids and hence need boundary layer parameterizations. Again, research experience with detailed PBL models can give these parameterizations.

6. Concluding Remarks

In this Introduction we have discussed some elementary aspects of the atmospheric boundary layer, and in the following sections we will pursue them in more detail. We shall not be able in this report to cover all important aspects of the subject, however. For example, we can touch only briefly on the marine boundary layer and on boundary layer observations, both important topics.

References

Batchelor, G.K., 1953: *The Theory of Homogeneous Turbulence*, Cambridge Univ. Press, Cambridge, England, 197 pp.

Blackadar, A.K. and H. Tennekes, 1968: Asymptotic similarity in neutral barotropic planetary boundary layers. *J. Atmos. Sci.*, 25, 1015-1020.

Businger, J.A. and S.P.S. Arya, 1974: Height of the mixed layer in the stably stratified planetary boundary layer. *Advances in Geophysics*, 18A, Academic Press, New York.

Clark, R.H., A.J. Dyer, R.R. Brook, D.G. Reid and A.J. Troup, 1971: The Wangara experiment: boundary-layer data. Paper No. 19, Div. of Meteorol. Physics, CSIRO, Australia.

Coantic, M. and B. Seguin, 1971: On the interaction of turbulent and radiative transfers in the surface layer. *Boundary-Layer Meteorol.*, 1, 245-263.

Csanady, G.T., 1974: Equilibrium theory of the planetary boundary layer with an inversion lid. *Boundary-Layer Meteorology*, 6, 63-79.

Deardorff, J.W., 1972: Numerical investigation of neutral and unstable planetary boundary layers. *J. Atmos. Sci.*, 29, 91-115.

Deardorff, J.W., 1973: An explanation of anomalously large Reynolds stresses within the convective planetary boundary layer. *J. Atmos. Sci.*, 30, 1070-1076.

Deardorff, J.W., 1976: On the entrainment rate of a stratocumulus-topped mixed Layer. *Quart. J. Roy. Meteorol. Soc.*, 102, 563-582.

Deardorff, J.W., 1974: Three-dimensional numerical study of turbulence in an entraining mixed layer. *Boundary-Layer Meteorol.*, 7, 199-226.

Delage, Y., 1974: A numerical study of the nocturnal atmospheric boundary layer. *Quart. J. Roy. Meteorol. Soc.*, 100, 351-364.

Funk, J.P., 1960: Measured radiative flux divergence near the ground at night. *Quart. J. Roy. Meteorol. Soc.*, 86, 382-389.

Gavrilov, A.S. and D.L. Laikhtman, 1973: Influence of radiative heat transfer on the conditions in the atmospheric surface layer. *Ozv. Atmos. and Oceanic Phys.* 9, 12-15.

Gent, P.R. and P.A. Taylor, 1976: A numerical model of the air flow above water waves. *J. Fluid Mech.*, 77, 105-128.

Haugen, D.A., ed., 1973: *Workshop on Micrometeorology*, AMS, Boston, Ma., 392 pp.

Hupfer, P., T. Foken and G. Panin, 1975: Existence and structure of the laiminar boundary layer of the atmosphere in the near-shore zone of the sea. *Zeitschrift für Meteorologie*, 25, 94-102.

Kaimal, J.C., J.C. Wyngaard, Y. Izumi and O.R. Coté, 1972: Spectral characteristics of surface-layer turbulence. *Quart. J. Roy. Meteorol. Soc.*, 98, 563-589.

Kaimal, J.C., J.C. Wyngaard, D.A. Haugen, O.R. Coté, Y. Izumi, S.J. Caughey and C.J. Readings, 1975: Turbulence structure in the convective boundary layer. *J. Atmos. Sci.*, 33, 2152-2169.

Kazanski, A.B. and A.S. Monin, 1961: Dynamic interaction between the atmosphere and the earth's surface. *Izv. ANSSSR, Ser. Geofiz.*, 5, 786-799.

Kolmogorov, A.N., 1941: The local structure of turbulence in the incompressible viscous fluid for very large Reynolds numbers. *Compt. Rend. Akad Nauk SSSR*, 30, 301-305.

Lilly, D.K., 1968: Models of cloud-topped mixed layers under a strong inversion. *Quart. J. Roy. Meteorol. Soc.*, 94, 292-309.

Lumley, J.L. and H.A. Panofsl. 1964: *The Structure of Atmospheric Turbulence.* Interscience-Wiley, New York, 239 pp.

Panofsky, H.A., H. Tennekes, D.H. Lenschow and J.C. Wyngaard, 1977: The characteristics of turbulent velocity components in the surface layer under convective conditions. *Boundary-Layer Meteorol.*, 11, 355-361.

Schlichting, H., 1960: *Boundary Layer Theory.* Mc-Graw-Hill, New York, 647 pp.

Tennekes, H. and J.L. Lumley, 1972: *A First Course in Turbulence.* MIT Press, Cambridge, Mass., 300 pp.

Weber, A.H., 1975: Atmospheric turbulence properties in the lowest 300 meters. EPA-600/4-75-004, National Technical Information Service, Springfield, Va.

Wippermann, F., 1973: *The planetary boundary layer of the atmosphere.* Deutschen Wetterdienst, Offenbach.

Wyngaard, J.C., 1973: On surface-layer turbulence. *Workshop on Micrometeorology,* (D.A. Haugen, ed.); AMS Boston, Mass., 101-149.

Wyngaard, J.C., 1975: Modeling the planetary boundary layer - extension to the stable case. *Boundary-Layer Meteorol., 9,* 441-460.

Wyngaard, J.C., S.P.S. Arya and O.R. Coté, 1974: Some aspects of the structure of convective planetary boundary layers. *J. Atmos. Sci., 31,* 747-754.

Wyngaard, J.C., O.R. Coté and K.S. Rao, 1974: Modeling the atmospheric boundary layer. *Advances in Geophysics, 18A,* Academic Press, New York, 193-211.

Wyngaard, J.C., W.T. Pennell, D.H. Lenschow and M.A. LeMone, 1977: The temperature-humidity covariance budget in the convective boundary layer. *J. Atmos. Sci.*

Yaglom, A.M., 1977: Comments on wind and temperature flux-profile relationships. *Boundary-Layer Meteor., 11,* 89-102.

Zeman, O. and H. Tennekes, 1977: Parameterization of the turbulent energy budget at the top of the daytime atmospheric boundary layer. *J. Atmos. Sci., 34,* 111-123.

Zilitinkevich, S.S., 1972: On the determination of the height of the Ekman boundary layer. *Boundary-Layer Meteor., 3,* 141-145.

CHAPTER 2 - EKMAN LAYER THEORIES

2.1 Introduction

In Chapter 1 an introduction to the physical aspects of the planetary boundary layer was presented. In this Chapter, the upper part of the PBL, the Ekman layer (EL) will be considered in more detail. The Ekman layer* is usually considered to extend from the top of the surface layer to the top of the PBL, commonly covering approximately 90% of the total PBL thickness. The basic characteristics of the EL, as opposed to the surface layer, are the variation of the turbulent fluxes and the turning of the wind vector with height. Because the wind vector as a function of height describes a curve similar to the logarithmic spiral, the EL is sometimes called the "spiral layer".

Within the EL the turbulent flux of momentum decreases from its surface value at z = 0 to near zero values at the top (in a manner that is complicated due to baroclinicity and stratification). Because the turbulent exchange coefficients increase to a maximum in the upper part of the EL the vertical gradients of most meteorological elements (e.g., wind speed, potential temperature, specific humidity) are not as steep as are usually observed in the surface layer. Also the amplitude of the diurnal variation of parameters which are strongly affected by the interaction with the underlying surface is reduced as compared with the surface layer and decreases with height. The diurnal variation also has an increasing phase delay with height.

The EL must be considered as the link between the surface layer, and the free atmosphere. The EL is characterized by the physical processes, such as turbulence, wind shear and clouds, which influence the meteorology of both the lower troposphere and the earth's surface. The divergence of the frictionally-induced ageostrophic flow within the PBL, which is concentrated mainly in the EL, causes the large-scale vertical motions which represent the main link between the PBL and the synoptic-scale dynamics. The EL will also have its own special pecularities such as low-level jets and characteristic thermal and moisture patterns. All of the above give rise to the special importance of the PBL and the EL in atmospheric physics and in fields of applied meteorology, such as aeronautical meteorology, acoustic and electromagnetic wave propagation and the dispersion of pollutants (some of these topics will be reviewed in Chapter 6).

---

* It should be noted that the term Ekman layer is sometimes used in the more restrictive sense of the PBL for stationary, neutral and horizontally homogeneous conditions. Also the EL is occasionally used for the whole of the PBL. In the context of this Technical Note the EL will be considered more generally as designated and is the same as the "outer layer" as opposed to the "inner", surface or wall layer, as used by some authors.

In comparison with the surface layer, our state of knowledge of the EL is very insufficient. The similarity theory of the EL (to be discussed in Section 2.2) was first introduced by Rossby (1932), Rossby and Montgomery (1935) and further developed by Kazanski and Monin (1960, 1961) and others. The similarity theory was originally developed for the neutral, barotropic, stationary and horizontally homogeneous PBL above a homogeneous surface. The results have recently been generalized to account for non-neutral and baroclinic conditions, with some attempt made to include special types of non-stationarity (diurnal changes). The effects of baroclinicity, inhomogeneity and non-stationarity are more pronounced in the EL (as discussed in Chapter 1) and, furthermore, the radiation processes play a significant role in the heat balance of the EL. For example, the radiative flux divergence may affect the vertical distribution of temperature which in turn influences the turbulence, wind and humidity profiles with a resulting feedback mechanism to the radiation fluxes. Therefore, the present similarity theory of the EL is much less valid and has many more restrictions than the similarity theory (Monin-Obukhov theory) of the surface layer.

Furthermore, there is a shortage of complete, high-quality data sets on the EL. Among the most comprehensive available are from O'Neill, Nebraska, U.S.A. (Lettau and Davidson, 1957) and Hay, New South Wales, Australia (Clarke et al., 1971) (N.B., these and other experiments are summarized in Chapter 5). Wind profile data up into the EL are available from Leipzig (Mildner, 1932) and Meppen, Federal Republic of Germany (Leykauf, 1974). More recently a comprehensive study was conducted in Minnesota, U.S.A. (Kaimal et al, 1975). Each of these was conducted over relatively uniform surfaces. PBL and EL data have also been collected as parts of major atmospheric experiments such as BOMEX, ATEX, JASIN, CAENEX, AMTEX and GATE (see Chapter 5 for further details).

Within the last decade there have been several articles reviewing the EL, such as: Zilitinkevich et al. (1967), Monin (1970), Zilitinkevich and Monin (1974), Tennekes (1974), de Moor (1975), de Moor and Andre (1975) and Counihan (1975); the latter one dealing with only the neutral PBL. Several monographs have also treated the EL, including those of Monin and Yaglom (1971), Laikhtman (1970), Zilitinkevich (1970), Brown (1974c) and the collected papers of Wippermann (1973). In addition, there are the very relevant proceedings of three symposia: Turbulence in the Boundary Layer of the Atmospheric, Moscow, 1973 (Izvestiya of Academy of Sciences, USSR, Atmospheric and Oceanic Physics, Vol. 10, Number 6, 1974; and GARP WGNE Report No. 5, 1974); JOC Study Conference on Parameterization, Leningrad, 1972 (GARP Publication Series, No. 8, 1972); and IUTAM-IUGG Conference on Turbulent Diffusion on Environmental Pollution, Charlottesville, Virginia, 1974 (Advances in Geophysics, Vols. 18A and 18B).

2.2 Flux-Profile Relationships in the Ekman Layer

The main topic of this Section is the description and modelling of the mean vertical profiles of wind, temperature and humidity and the distributions of characteristic turbulence parameters such as turbulent fluctuations, eddy, viscosity, conductivity and diffusivity. We will start with the usual relationship between the vertical flux and the gradient (see Equations 1.3.4).

$$F = \rho K \frac{\partial S}{\partial z} \qquad (2.2.1)$$

where F is any vertical flux (e.g. τ, H, M) and S is the corresponding mean quantity (e.g. U, θ, Q). As has been pointed out (Ertel, 1932, 1937; Monin and Yaglom, 1971; Bernhardt, 1970a; and in Chapter 1 of this Technical Note) this is a coarse approximation to the real PBL behaviour and is also tensorially incorrect. Nevertheless this relationship has been used with a certain amount of success by numerous researchers.

2.2.1 Ad hoc Approaches for Turbulent Transfer Coefficients

The models to be described in this Section use the a priori approach of specifiying the vertical profile of eddy viscosity, K_m, within the PBL. Ekman (1905, 1906), for his theory of wind-driven ocean currents, and Akerblom (1908), Exner (1912) and Taylor (1915) for applications in the atmosphere, assumed that K_m = constant for all z. This led to the Ekman spirals which were only a very crude approximation of the real wind profile. Subsequent developments, using K_m's which depended on z yielded much better agreements with observations. Some of the models of this type are listed in Appendix I.

Although these ad hoc approaches are unsatisfactory from a physical point of view (since they are based neither on a closed system of equations nor on similarity theory) the results obtained may be good enough for many practical purposes. In many cases the $K_m(z)$'s lead to rather simple expressions for the wind profile. Even the simple single-layer power-law approach leads to an expression for the geostrophic drag coefficient which is equivalent to that given by Swinbank (1974) as a good approximation to the resistance law (see Section 2.3.1). These K_m's are frequently used in PBL numerical models as will be seen in Chapter 4.

The effects of baroclinicity may also be taken into account using the ad hoc approach with the occurring parameters being determined from observations by empirical means.

2.2.2 Closed Semi-Empirical EL Equation Systems

In this Section the approach to K(z) using closed boundary-layer equations is described (for further details, see, e.g. Laikhtman, 1970; Zilitinkevich, 1970; Wippermann, 1973a). Only the barotropic, stationary and horizontally homogeneous boundary layer will be considered. In this type of K-theory the eddy viscosity K_m (and the eddy conductivity and diffusibity in an analogous way) is related to the mixing length ℓ (also called the length scale of turbulence) either by the Prandtl approach,

$$K_m = \ell^2 \left|\frac{\partial \vec{V}}{\partial z}\right| \qquad (2.2.2)$$

with V being the horizontal wind vector or by the Kolmogorov relation

$$K_m = \ell b \tag{2.2.3}$$

with b^2 being the kinetic energy of turbulent motion per unit mass. In either case the formulation to determine the mixing length is the key problem in closing the boundary-layer equation system. Wippermann (1973a) has discussed the consistency between these approaches and Rossby similarity (Section 2.3).

The mixing length can be either prescribed as a specified function of height or can be related to other dependent variables. Examples of the former type are the approaches of Blackadar (1962):

$$\ell = \frac{kz}{1 + kz/0.00027G/f'} \tag{2.2.4}$$

Lettau (1962):

$$\ell = \frac{kz/h}{1 + k(z/h)^{5/4}} \tag{2.2.5}$$

where h is the scale height of the PBL and Ruzin et al. (1963):

$$\ell = \ell_\infty \left(1 - e^{-kz/\ell_\infty}\right) \tag{2.2.6}$$

for neutrally stratified boundary layers.

An example of the second type is:

$$\ell = kc^{1/4} \frac{\psi}{\partial\psi/\partial z} \tag{2.2.7}$$

$$\psi = \left[\left(\frac{\partial U}{\partial z}\right)^2 + \left(\frac{\partial V}{\partial z}\right)^2 - \alpha_H \frac{g}{\Theta}\frac{\partial \Theta}{\partial z} + \frac{\alpha_b}{K_m}\frac{\partial}{\partial z} K_m \frac{\partial b^2}{\partial z}\right]^{1/2} \tag{2.2.8}$$

where c is a non-dimensional constant; α_b is the ratio of the turbulent exchange coefficients for turbulent kinetic energy and momentum; and α_H is the ratio of the turbulent exchange coefficients for heat and momentum. This approach was introduced by Bobyleva et al. (1965) as a generalization of the earlier approaches by Karman (1930), Rossby (1932), Monin (1950) and Zilitinkevich and Laikhtman (1965). Equation (2.2.7) is combined with (2.2.3), the steady-state equation of turbulent kinetic energy $(\partial b^2/\partial t = 0)$, i.e.

$$\varepsilon = K_m \left[\left(\frac{\partial U}{\partial z}\right)^2 + \left(\frac{\partial V}{\partial z}\right)^2 - \alpha_H \frac{g}{\Theta}\frac{\partial \Theta}{\partial z} \right] + \alpha_b \frac{\partial}{\partial z} K_m \frac{\partial b^2}{\partial z} \qquad (2.2.9)$$

and a relationship for the turbulent kinetic energy.

$$\varepsilon = cb^2 \ell^{-1} \qquad (2.2.10)$$

to obtain a closed system of two equations of motion and the equation (2.2.3), (2.2.7), (2.2.9) and (2.2.10) with 6 unknown variables $U(z)$, $V(z)$, $K_m(z)$, $\ell(z)$, $b^2(z)$ and $\varepsilon(z)$ if α_H and α_b are given and the vertical profile of potential temperature or turbulent heat flux is prescribed. If the profile of heat flux or potential temperature is not given then the equation of heat balance can be used as a seventh equation.

The constant c in (2.2.7) has been shown in laboratory measurements to be,

$$c = \lim_{z \to 0} u_*^4/b^4 \simeq 0.046 \qquad (2.2.11)$$

Wippermann (1971) has used a different approach; relating the mixing length with the turbulent shearing stress by the relation:

$$\ell = \ell_\infty - (\ell_\infty - \ell_o) \left(\frac{\tau}{\tau_o}\right)^p \qquad (2.2.12)$$

with the exponent p is given by:

$$p = \frac{k\tau_o}{(\partial\tau/\partial z)_o (\ell_\infty - \ell_o)} \qquad (2.2.13)$$

when the subscript o refers to the surface values. The asymptotic value, ℓ_∞, non-dimensionalized by the scaling height of the PBL $\lambda = ku_*/f$ as a function of the stratification parameter μ_λ (see Section 2.3) was determined from empirical data to be (Wippermann, 1971):

$$\mu_\lambda = \lambda/L = -44.2 - 24.4 \log (\ell_\infty/\lambda)$$

while ℓ_o/λ is given by the relationship:

$$\ell_o/\lambda = kK_{mo}/\lambda^2 f = kz_o/\lambda = 1/C_g Ro$$

reflecting the effect of the Rossby number, Ro, and the geostrophic drag coefficient, C_g, (see Section 2.3). Then the two equations of motion with (2.2.2) and (2.2.12) constitute a closed system of equations for the four unknowns: $U(z)$, $V(z)$, $K_m(z)$ and $\ell(z)$.

Using the mixing length approach several authors have calculated vertical wind profiles in the PBL (usually non-dimensional) and also vertical profiles of eddy viscosity, stress, energy dissipation rate, etc., by numerical integration of a closed set equations with appropriate boundary conditions. Among these are: Blackadar (1962), Lettau (1962), Bobyleva et al. (1965), Yamamoto et al. (1968), Kurosaki (1970), Zilitinkevich et al. (1967), Zilitinkevich (1970, Section 2.3.2), Laikhtman (1970, Chapter 2 and Appendix), Wippermann (1972c; 1973, pp. 95-110) and Wippermann (1975).

In these semi-empirical PBL models the mixing lengths are not defined precisely in terms of physical theory but usually are based on some intuitive sense. The vertical heat flux, which has been specified in some of them, should, of course, be computed from the heat balance equation, including the radiation flux divergences. Some extensions of the models to baroclinic, nonstationary and/or inhomogeneous conditions are outlined briefly in Section 2.4.

2.3 The Similarity Theory and Resistance Laws of the Ekman Layer

2.3.1 The Ekman Layer Similarity Theory

The Ekman layer similarity theory (Zilitinkevich and Monin, 1974) gives the departure of the boundary layer wind from the geostrophic wind in terms of the surface friction velocity, u_* and universal functions F of non-dimensional height, z/h, (where h is the depth of the PBL) and stability parameter, $\mu = h/L$;

$$U(z) - G\cos\alpha = \frac{u_*}{k} F_U\left(\frac{z}{h}, \mu\right)$$

$$V(z) - G\sin\alpha = \frac{u_*}{k} F_V\left(\frac{z}{h}, \mu\right) \text{ sign } f \qquad (2.3.1)$$

G is the geostrophic wind speed and α is the angle between the geostrophic wind vector and the surface wind (assumed to be in the x direction so that U(0) = U and V(0) = 0). The sign f accounts for the differences between the Northern and Southern Hemispheres. These equations are referred to as Rossby Similarity or the auto-model hypothesis of the PBL. The profiles are also called the "wind-defect profiles".

By analogy, some other non-dimensional parameters such as: wind shear, $(k^2/f)\partial\vec{V}/\partial z$, shear stress, τ/τ_0, eddy diffusivity, K_m/ku_*h, can be regarded as universal functions of z/h and μ. The profiles of potential temperature and humidity may be expressed as:

$$\Theta(z) - \Theta(h) = \frac{\theta_*}{k} F_\Theta\left(z/h, \mu\right) \qquad (2.3.2)$$

and

$$Q(z) - Q(h) = \frac{q_*}{k} F_Q \left(\frac{z}{h}, \mu\right) \qquad (2.3.3)$$

where $q_* = M/\rho u_*$. Corresponding equations can be written for temperature and humidity fluctuation parameters, eddy conductivity and diffusivity, etc.

All the universal functions introduced in this theory have to be determined either empirically from observations or theoretically by the aid of closed systems of equations for the turbulent PBL. However, as yet, there is not complete agreement on the appropriate scaling height for all types of PBL stratification. For neutral and perhaps near neutral stratification the scaling height is apparently $h = \gamma u_*/f$, in accordance with dimensional analysis considerations with γ between 0.25 and 0.6 (most often nearer the lower limit). Zilitinkevich (1972a,b) has generalized this expression for the stationary and horizontally homogeneous PBL:

$$h = \frac{u_*}{f} \gamma \left(\mu_\lambda\right) \qquad (2.3.4)$$

where $\mu_\lambda = \lambda/L = ku_*/fL$. This means that the arguments of F_U, F_V, F_Q etc. involving the scale height h are themselves functions of λ and μ_λ. Therefore the dimensionless velocity, temperature and moisture defect profiles only depend on z/λ and μ_λ, (Kazanski and Monin, 1960). Asymptotic expressions for the universal functions γ (μ_λ) in (2.3.4) (Zilitinkevich, 1972a,b) are for strongly stable $(\mu >> 1,\ \gamma\ (\mu_\lambda) \sim |\mu|^{-1/2})$ and very unstable $(\mu << -1,\ \gamma\ (\mu_\lambda) \sim |\mu|^{1/2})$ stratifications.

However, for unstable conditions the numerical models of Deardorff (1970a, 1972b), Wyngaard et al. (1974a,b), and Sundarajan (1975) and the postulates of Carson (1973) and others suggest that the depth of the convective PBL is independent of λ and should be characterized by the height z_i of the lowest inversion base (see Sections 1.2.1, 1.3.2.1). In this case the universal function should depend on z/z_i and $\mu_i = z_i/L$ instead of z/λ and μ_λ. Moreover, the friction velocity may not be the appropriate velocity scale and should be replaced by w_* (1.3.22). Similarly θ_* and q_* are redefined to use w_* instead of u_* in their definitions. The k-theory concept becomes invalid under such conditions because the buoyancy induced motions transfer considerable momentum and heat even though the mean gradients of wind and temperature are near zero and the geostrophic adjustment is reached only above the inversion (e.g. Deardorff, 1973, 1974a,b; Arya, 1975b).

The analyses of the Wangara data by Clarke (1970a) and Clarke and Hess (1973) do not suggest the use of z_i as the scaling height but apparently prefer the use of λ for all stability classes. In addition, Kukharets and Tsvang (1976) found that the profiles of non-dimensional heat flux, potential temperature defect, and structure characteristics of two-point temperature fluctuations in the unstable PBL may be plotted as functions of dimensionless height and a stratification parameter. The height at which the turbulent heat flux vanished or was a minimum was used as the scaling height. This height was found to be closely correlated with $h_c = (g/T_o\ Q_o/f^3)^{1/2}$, as given by Zilitinkevich, and is equivalent to the asymptotic behaviour of $\gamma(\mu_\lambda)$ referred to above.

The question of how to determine the appropriate top of the PBL or EL from observations or in numerical or semi-empirical models has been discussed by Hanna (1969), Clarke (1971) and Etling and Wippermann (1975). There is still a scarcity of good data with which to attempt to determine the shape of the universal functions. Consequently the issue cannot be considered as being settled.

As discussed in Chapter 1 there is evidence that for unstable PBL's the height to which the turbulence and convection will extend will be limited to the height of the lowest inversion base. Schneider-Carius (1953) had earlier discussed the influence of the capping inversion layer, also called the peplopause, on the boundary layer.

For the real EL the effects of baroclinicity, nonstationarity and horizontal inhomogeneity are very important. EL theories need to be developed to account for these effects as well as for the radiation processes that affect the PBL energy balance. There is at present a serious need for more high quality observational data to test out similarity and other theories as well as for verification studies on numerical models.

2.3.2 Resistance Law Formulations

It should be noted that the EL similarity theory does not include the direct influence of surface roughness, z_o, and hence is only valid for $h/z_o \gg 1$ and $z/z_o \gg 1$. If the scaling height h is assumed proportional to u_*/f then the former is equivalent to:

$$C_g R_o \gg 1 \qquad (2.3.5)$$

where $C_g = u_*/G$, the geostrophic drag coefficient, G is the geostrophic wind speed, and $R_o = G/fz_o$, the surface Rossby number. By assuming that the ranges of the Rossby number similarity equations (2.3.1, 2.3.2, 2.3.3) and of the surface layer logarithmic profiles overlap it was shown (see Monin and Zilitinkevich, 1974; Zilitinkevich and Deardorff, 1974) by asymptotic matching that:

$$\frac{1}{C_g} = \frac{1}{k} \left\{ \left(\ln \frac{h}{z_o} - b(\mu) \right)^2 + a^2(\mu) \right\}^{1/2} \qquad (2.3.6)$$

$$\sin \alpha = - \frac{C_g\, a(\mu)}{k} \operatorname{sign} f \qquad (2.3.7)$$

$$\frac{\Theta(h) - \Theta(o)}{\theta_*} = \frac{1}{k\alpha_H{}^o}\left[\ln \frac{h}{z_o} - c(\mu)\right] \qquad (2.3.8)$$

$$\frac{Q(h) - Q(o)}{q_*} = \frac{1}{k\alpha_D{}^o}\left[\ln \frac{h}{z_o} - d(\mu)\right] \qquad (2.3.9)$$

are the appropriate resistance laws for the stationary, horizontally homogeneous, barotropic boundary layer. The terms $1/\alpha^o{}_H$ and $1/\alpha^o{}_D$ are the turbulent Prandtl and Schmidt numbers in the range of the logarithmic profiles near the surface. It should be noted that the a and b in (2.3.6) and (2.3.7) are interchanged in meaning in some articles.

The equations (2.3.6-2.3.9) permit, in principle, the calculation of the surface stress and the surface fluxes of sensible and latent heat from the external boundary layer parameters, z_o, G, $\Theta(h)$, $\Theta(o)$, Q(h), Q(o). However, it is required that:

(i) the scaling height be determined from observations or be specified in terms other external parameters or from additional equations;

(ii) the universal functions a, b, c and d as functions of μ be known and that $\mu = h/L$ = function of u_*, T_* and q_* can be solved for iteratively;

(iii) $\alpha_H{}^o$ and $\alpha_D{}^o$ be known.

As yet, as will be seen, our knowledge of this is imperfect.

If the length scale λ is used for h then the equations are usually written as:

$$\ln C_g R_o = B(\mu_\lambda) \sqrt{\frac{k^2}{C_g{}^2} - A^2(\mu_\lambda)} \qquad (2.3.10)$$

$$\sin \alpha = \frac{C_g}{k} A(\mu_\lambda) \text{ sign } f \qquad (2.3.11)$$

$$\frac{\Theta(h) - \Theta(o)}{\theta_*} = \frac{1}{k\alpha_H{}^o}\left[\ln C_g R_o - C(\mu_\lambda)\right] \qquad (2.3.12)$$

$$\frac{Q(h) - Q(o)}{q_*} = \frac{1}{k\alpha_D{}^o}\left[\ln C_g R_o - D(\mu_\lambda)\right] \qquad (2.3.13)$$

as was deduced by Kazanski and Monin (1961), Blackadar (1967), Blackadar and Tennekes (1968), Csanady (1967, 1972), Gill (1968), Zilitinkevich (1967a, 1970) and others.

The function a, b, c, d, or A, B, C, D or F_U, F_V, F_Q are all some-what corresponding and must be determined from observations or from models. Zilitinkevich and Chalikov (1968a, from O'Neill data), Clarke (1970a, Kerang and Hay data), Clarke and Hess (1974, Wangara data) and Arya (1975a, re-analysis of Wangara data) have all obtained values of A, B, C and D. Further values were added by Soloviev (1973, Obninsk data) and Dubov and Kucherov (1972, Pacific expedition data), while Wippermann (1970, 1972b) summarized A and B values (which he calls N and M) from various sources and also developed empirical analytical formulae for the functions.

In most of the above cited papers, data on C_g and α were also presented. Other data on these parameters have recently been presented by Hasse and Dunckel (1974, North Sea Observations), Bernhardt (1973, 1974b, 1975, routine rawinsonde data from Lindenberg, GDR), Brown (1974a,b) and La Valle and Di Girolamo (1975). Deacon (1973) and Rüffer (1976) have summarized the geostrophic drag coefficients; many of which are reproduced in Appendix II.

Arya (1975) and Melgarejo and Deardorff (1974) reanalysed the Wangara wind and potential temperature profiles and determined the a, b, c functions in terms of observed boundary layer heights.

Based on model calculations Zilitinkevich (1970, Sec. 2.3.2.2) has presented analytic expressions for μ_λ, while Wippermann (1972b) and Arya (1975a) have developed empirical interpolation formulae to fit observational data. Zilitinkevich (1975) has deduced asymptotic expressions for $A(\mu_\lambda)$, $B(\mu_\lambda)$ and $C(\mu_\lambda)$, as well as for a(h/L) and b(h/L) for strongly stable and strongly unstable conditions and also for slightly unstable conditions of shear convection (Zilitinkevish, 1973). Businger and Arya (1974) recently developed a model for the stably stratified PBL.

In spite of the considerable scatter of the data from various sources there is some agreement on general characteristics. The functions are nearly constant for moderate to large instability with A and a increasing and B, C, D, b, c, d decreasing as the stability becomes neutral. This is followed by a rapid increase (A, a) or decreases (B, C, D, b, c, d) as stability increases further. The data for stable conditions show a particularly high degree of scatter. Values of A, B, C, or a, b, c from both model experiments (see Section 2.2) and observations are listed in Table 2.1.

2.3.3 The Parameterization of the Boundary Layer

The relationships (2.3.6-2.3.9) and (2.3.10-2.3.13) can be used to develop parameterization schemes for the boundary layer. For example, in numerical models with coarse vertical resolution which treat the PBL as a whole, these parameterizations will give the fluxes directly. Since the a, b, c, d or A, B, C, D's are all functions of μ or μ_λ which depend on u_* and θ_*, through L, it is necessary to solve the equations iteratively. An alternate approach is to express μ or μ_λ in terms of the external parameters. Combining (2.3.6) and (2.3.8) gives:

$$S = \beta h \frac{\left(\Theta(h) - \Theta(0)\right)}{G^2} = \frac{\mu}{\alpha_H^0} \left[\frac{\ln \frac{h}{z_o} - c(\mu)}{\left(\ln \frac{h}{z_o} - b(\mu)\right)^2 + a^2(\mu)} \right] \qquad (2.3.14)$$

Table 2.1 - Values of the Universal Functions in Resistance Laws

		$A(\mu_\lambda)$	$B(\mu_\lambda)$	$C(\mu_\lambda)$
a.	Typical Values			
	unstable $(\mu_\lambda < 0$	$0 \leftrightarrow 2$	$3 \leftrightarrow 5$	$3 \leftrightarrow 8$
	near neutral $(\mu_\lambda \approx 0)$	$4 \leftrightarrow 6$	$0 \leftrightarrow 2$	$-3 \leftrightarrow 3$
	stable $(\mu_\lambda > 0)$	$10 \leftrightarrow 20$	$-40 \leftrightarrow -10$	$-60 \leftrightarrow -20$
b.	Model Experiments (neutral)			
	Monin (1950)	1.5	5.0	
	Blackadar (1962)	6.7	1.8	
	Lettau (1962)	4.6	0.9	
	Bobyleva et al. (1965)	2.0	2.2	
	Wipperman (1972c)	4.5	1.1	
	Wipperman (1975)	4.6	0.9	-4.3
	Fiedler (1972)	4.4	0.8	
c	Empirical Results (near neutral)			
	Wippermann (1970)	4.5	0.9	
	Deacon (1973)	4.7 ± 0.15	1.9 ± 0.35	
	Clarke and Hess (1974)	4.3 ± 0.7	1.1 ± 0.5	
	Melgarejo and Deardorff (1974a,b)	≈ 4.5 (a ≈ 4.5)	≈ 0 (b ≈ -1.5)	≈ 3 (c ≈ 2)
	Arya (1974a)	4.1	1.0	3.0
	Yordanov (1975)	4.4	1.0	

with S, the bulk Richardson number, as an external parameter. Alternatively with λ as the length scale (2.3.10) and (2.3.13) combine to give:

$$S_\lambda = \beta \frac{\{\Theta(h) - \Theta(0)\}}{Gf} = \frac{\mu_\lambda}{k^2 \alpha^0_H} C_g \left[\ln R_o C_g - C(\mu_\lambda) \right] \qquad (2.3.15)$$

Zilitinkevich and Chalikov (1968b), Chalikov (1968), Clarke (1970b), Arya (1975a) and Wippermann (1975) have presented nomograms to determine the internal stability parameters μ_λ as well as the geostrophic drag coefficient C_g, the turning angle α, and surface non-dimensional heat and moisture fluxes from Ro and S_λ. A nomogram to evaluate μ_λ from isolation and G was produced by Fieldler and Panofsky (1972). Deardorff (1972) has recommended a different approach for parameterization of the PBL in numerical models. His method uses the stability and time dependent heights. Melgarejo and Deardorff (1974a,b) published nomograms for the determination of $\mu = h/L$, C_g, α and the fluxes from the bulk Richardson number and h/z_o. The boundary-layer height h and the wind at z = h are used and the nomograms are based on the Wangara data. Another approach for PBL parameterization based upon integral characteristics of the PBL was proposed by Ordanovich and Ivanov (1974).

To determine the Ro it is necessary to first know the surface roughness z_o. Over land the effects of surface elevations may give rise to an additional contribution of form drag to the shear drag and hence the effective z_o may differ from the locally determined z_o (e.g. Arya, 1973). The effective roughness concept was formulated by Fiedler and Panofsky (1972) while Wieringa (1974) has proposed a method of determining an approximate roughness length from wind fluctuation statistics.

Over the sea the roughness length will depend on the state of the sea surface waves. A variety of relationships to determine z_o have been proposed by, e.g. Monin and Zilitinkevich (1967, Zilitinkevich (1967b), Bernhardt (1970a, 1970b), Dubov and Orlenko (1970), Dubov (1975), Wippermann (1972a, 1974), Kitaigorodsky (1968, 1970) and Brocks and Krugermeyer (1972). The most well known is that of Charnock (1955), $z_o = a u_*^2/g$. Wieringa (1974) has pointed out that form drag on waves may result in a longer effective drag coefficient (and hence z_o) than is apparent from point measurement.

There is still a strong need for observational data for the better determination of the coefficients in these parameterization schemes. Also the systematic verification of different parameterization schemes in numerical models, as done by Delsol et al. (1971), Everson and Davies (1972, Fischer et al. (1973), Boye (1974) and Heise (1975), should be encouraged.

2.4 The Effects of Baroclinicity, Nonstationarity and Horizontal Inhomogeneity

2.4.1 The Effects of Baroclinicity

Baroclinicity, the variation of the geostrophic wind vector with respect to height, influences strongly the wind profile within the PBL as well as the deviation of the surface wind from the direction of the isobars, the geostrophic drag coefficient and other properties. Modelling of the effects of baroclinicity leading to the modifications of the Ekman spirals (K_m assumed to be constant with height) was done by Mahrt and Schwerdtfeger (1970) and MacKay (1971). Hess (1973) and Yordanov and Wippermann (1972) extended the Rossby similarity theory to baroclinic PBL by use of non-dimensional baroclinicity parameters which expressed the components of the geostrophic wind shear and were assumed to be independent of height. Then the parameters, A, B, etc. became functions either of the internal stratification parameter and the two internal baroclinicity parameters or of the external stratification parameter, two external baroclinicity parameters and the Rossby number. Wippermann (1972d) calculated these functions by use of his model and presented tables and nomograms. Yordanov (1974a,b,c,d) also deduced the function describing the influence of baroclinicity on the dimensionless wind and stress profiles in the PBL. Yordanov's evaluation of the dependence of the parameters A, B on baroclinicity, however, is not in agreement with the experimental results from the Wangara data (Clarke and Hess, 1974) which are approximately validated by Clarke's (1975) theoretical considerations.

Other theoretical approaches to modelling the influence of baroclinicity on the PBL structure have been developed by Fiedler (1972), who applied a mixing-length approach to obtain A, B as functions of stratification and baroclinicity; Voltsinger et al. (1973), who accounted for the variation of the pressure gradient with height as well as the horizontal temperature gradient in the heat balance equation; and Yordanov (1975), who used a two-layer PBL model (Yordanov, 1974e) which provided the resistance law as a function of stability and baroclinicity (see also Yordanov and Dimitrov, 1974, who compared the approaches of Zilitinkevich, Chalikov and Blackadar). Venkatesh and Csanady (1974) also used a simple two-layered baroclinic PBL model and applied it to the Wangara data. This approach was criticized by Arya (1975b).

Arya and Wyngaard (1975) modelled the baroclinicity effect on wind profiles and on the geostrophic drag coefficient for convective PBL's; the results were found to be in reasonable agreement with the analyses of Hoxit (1974) and Brümmer et al. (1974). Clarke and Hess (1975), however, analysed climatological pressure, wind and temperature fields over the ocean and concluded that the dependence of latitude of α, the geostrophic departure angle, and the ratio, U/G, may be explained satisfactorily by Rossby similarity including the effects of baroclinicity after Clarke and Hess (1974). The formulation of Arya and Wyngaard (1975), based on the well-mixed convective boundary layer assumption, did not agree with the observations as well. Laikhtman and Popov (1976) explained the behaviour of α with decreasing latitude (which was discussed by Arya and Wyngaard and Clarke and Hess) without regard to baroclinicity but referring to an extended similarity approach taking into account the vertical temperature gradient at the top of the PBL as a parameter in addition to Ro and μ_λ.

Hence, it must be concluded that, although the effects of baroclinicity may be of great importance and modelling of the effects is under development, widely accepted results have not yet been obtained.

2.4.2 The Effects of Nonstationarity

To date there have not been very many modelling studies on the effects of nonstationarity on PBL structure. Some examples where semi-empirical theories were used, are Vager and Zilitinkevich (1968, see also Zilitinkevich, 1970, Section 2.4), Lykosov (1972) and Wippermann et al. (1973) who studies the diurnal variation of temperature (the first two) or of heat flux (Wippermann) at the ground or assumed periodic as well as non-periodic changes in the gestrophic wind (Lykosov). These authors computed PBL characteristics as functions of the stability and of time and compared the results with those from calculations with steady-state conditions.

More recently Djolov (1974; see also Djolov and Syrakov, 1975), using a mixing-length hypothesis described the variations in PBL structure caused by temperature and moisture changes at the surface and by temperature, moisture and wind variations in the free atmosphere. The results may be compared with observed time-dependent PBL parameters, as presented by Buajitti and Blackadar (1957), Clarke (1974) and Clarke and Hess (1974). These results confirmed the "hysteresis" effects found on the calculations by Vager and Zilitinkevich and Wippermann et al. and qualitatively in the Wangara data. More recently Orlenko and Shklyarevich (1975) found considerable disagreement between the observed diurnal variation of C_g and α and the results of Vager and Zilitinkevich's model.

Wippermann (1973b) concluded on the basis on his numerical experiments that the combined effects of thermal stratification, baroclinicity and non-stationarity were responsible for the occurrence of low-level jet streams. In this connection, the inertial oscillations resulting from the diurnal change of thermal stratification and of eddy viscosity also contribute to the diurnal oscillatory wind changes in the upper part of the PBL and to the night-time lower troposphere wind speed maxima, as demonstrated by Blackadar (1957), Buajitti and Blackadar (1957), Bonner and Paegle (1970) and Hoxit (1975). Bonner and Paegle also found a significant diurnal variation in the surface geostrophic wind over the south central U.S.A. This effect complicates the relationships between the time-dependent wind and temperature profiles which restricts the feasibility of parameterization and limits the applicability of the similarity in the form as developed so far.

A further complicating feature of the nonstationarity of the PBL is the behaviour of inversions commonly capping the convective PBL's (but not only these) which are as a rule composed of a strongly unstable surface layer, a nearly neutral layer and a pronounced stable layer at the top, frequently located above a layer of dense haze or stratiform clouds. Ball (1969), Lilly (1968), Deardorff et al. (1969), Tennekes (1973), Carson (1973), Stull (1973) Gossard and Moninger (1975) and others have contributed to the understanding of relevant physical mechanisms and the modelling of this type of PBL (see also Sections 1.2.1, 1.3.2.1).

Within the framework of the similarity theory Deardorff (1972a), Zilitinkevich and Deardorff (1974) and Zilitinkevich (1975) considered PBL's of time-dependent height, deducing prognostic equations for the thickness of the PBL and proposing methods for the parameterization of such PBL's.

The release of boundary-layer content into the free atmosphere also appears to be a nonstationary process. Some ideas regarding these processes were described by Priestley (1967) and recently Bernhardt (1976) outlined four mechanisms for the mesoscale and synoptic scale release, namely: residual turbulence, mesoscale vertical motions, macroscale vertical motions and gravity waves.

2.4.3 The Effects of Horizontal Inhomogeneity

Perhaps the main defect of boundary-layer theories is the neglect of the effects of horizontal inhomogeneity within the PBL and at the underlying surface. Most theories and observational studies have dealt with the ideal homogeneous surface which is not typical of the earth's surface. The model experiments of Vorob'eva et al. (1968) demonstrated the significant role of the horizontal inhomogeneity of the PBL for the dynamics of large-scale processes. These inhomogeneities are essential for air-mass transformations and are important for mesoscale predictions. Some aspects of the PBL over inhomogeneous underlying surface are considered by Laikhtman (1970, Chapter 4), Estoque 1973, Sections 6.4, 6.5). Lykosov and Gutman (1972) and Speranskii et al. (1975) investigated theoretically the special problem of the PBL over a sloping and over a curvelinear underlying surface. Taylor (1977a,b) considered the PBL flow above gentle topography.

The behaviour of the PBL over changing surface roughness has been considered by Taylor (1969), Rao et al. (1974), Shir (1972), Vager and Nadezhina (1975a), Lo (1977) and others. Vager and Nadezhina (1973, 1975) also investigated the effects of a change in surface moisture and albedo upon the structure of the PBL.

Generally only the special cases of single abrupt change inhomogeneities have been studied. Few studies exist on boundary layer flow over typical surface types with irregularly distributed elevations and other inhomogeneities and further investigations are needed.

References

Akerblom, F., 1908: Recherches sur les courants les plus bas de l'atmosphère au-dessus de Paris. *Nova Acta Soc. Sci. Ups.* 2, Ser. 4, Nr. 2, 1-45.

Arya, S.P.S., 1973: Contributions of form drag on pressure ridges to the air stress on Arctic ice. *J. Geophys. Res.*, 78, 7092-7099.

Arya, S.P.S., 1975a: Geostrophic drag and heat transfer relations for the atmospheric boundary layer. *Quart. J. Roy. Meteor. Soc.*, 101, 147-161.

Arya, S.P.S., 1975b: Comments on "a baroclinic planetary boundary-layer model and its applications to the Wangara data". *Boundary-Layer Meteor.*, 9, 123-126.

Arya, S.P.S. and J.C. Wyngaard, 1975: Effect of baroclinicity on wind profiles and the geostrophic drag law for the convective boundary layer. *J. Atmos. Sci.*, 32, 767-778.

Ball, F.K., 1960: Control of inversion height by surface heating. *Quart. J. Roy. Meteor. Soc.*, 86, 483-494.

Bernhardt, K., 1970a: Eine Erweiterung des halbempirischen Austauschansatzes. *Gerl. Beitr. Geophys.*, 79, 345-362.

Bernhardt, K., 1970b: Der ageostrophische Massenfluß in der Bodenreibungsschicht bei beschleunigungsfreier Strömung. *Z. Meteor.* 21, 259-279.

Bernhardt, K., 1973: Zur Parametrisierung von Turbulenzreibungseffekten in der planetarischen Grenzschicht. *Gerl. Beitr. Geophys.* 82, 291-298.

Bernhardt, K., 1974a: Über den Einfluß der planetarischen Grenzschicht über Land und Meer auf die großräumige atmosphärische Dynamik. *Z. Meteor.*, 24, 16-22.

Bernhardt, K., 1974b: Preliminary results of the determination of the geostrophic drag coefficient from rawinsonde data. (In Russian). *Izvestiya Acad. Sci., USSR, Atmosph. Ocean Phys.*, 10, 665-668.

Bernhardt, K., 1975: Some characteristics of the dynamic air-surface interaction in Central Europe. *Z. Meteor.*, 25, 63-68.

Bernhardt, K., 1976: Mesoscale and synoptic scale release. WMO CAS Working Group on Atmosph. Bound.-Layer Probl., First Session, Potsdam, 20-24 April 1976, Final Rep., Appendix D, 4 pp.

Blackadar, A.K., 1962: The vertical distribution of wind and turbulent exchange in a neutral atmosphere. *J. Geophys. Res.*, 67, 3095-3102.

Blackadar, A.K., 1967: External parameters of the wind flow in the barotropic boundary layer of the atmosphere. Report of the Study Conference held at Stockholm on the Global Atmospheric Research Programme (GARP) Appendix IV, 11 pp.

Blackadar, A.K. and H. Tennekes, 1968: Asymptotic similarity in neutral barotropic boundary layer. *J. Atmos. Sci.*, 25, 1015-1020.

Bobyleva, I.M., S.S. Zilitinkevich and D.L. Laikhtman, 1965: The regime of turbulence in a thermically stratified planetary boundary layer of the atmosphere. Atmosph. Turbulence and Radiowave Prop., Proc. Internat. Collog. Moscow, 15-22 June, Nauka, Moscow, 179-190.

Bonner, W.D. and J. Paegle, 1970: Diurnal variations in boundary layer winds over the south-central United States in summer. *Mon. Weath. Rev.*, 98, 735-744.

Boye, H., 1974: Numerische Rechnungen unter Berücksichtigung der Ekman-Schicht in einem baroklinen Vorhersagemodell. *Hamburger Geophys. Einzelschr.* 20, 103 S.

Brock, K. and L. Krügermeyer, 1972: The hydrodynamic roughness of the sea surface. Studies in Physical Oceanography (A.L. Gordon, ed.), Gordon and Breach, New York, 75-92.

Brown, R.A., 1974c: Planetary Boundary Layer Modelling. London, Adam Hilger Ltd., 150 pp.

Brümmer, B.E., E. Augstein and H. Riehl, 1974: On the low-level wind structure in the Atlantic trade. *Quart. J. Roy. Meteor. Soc.*, 100, 109-121.

Buajitti, K. and A.K. Blackadar, 1957: Theoretical studies of diurnal wind-structure variations in the planetary boundary layer. *Quart. J. Roy. Meteor. Soc.*, 83, 486-500.

Businger, J.A. and S.P.S. Arya, 1974: Height of the mixed layer in the stably stratified planetary boundary layer. *Turb. Diff. Environm. Pollut., Adv. Geophys.* 18a, New York, 73-92.

Carson, D.J., 1973: The development of a dry inversion-capped unstable boundary layer. *Quart. J. Roy. Meteor. Soc.*, 99, 450-467.

Chalikov, D.V., 1968: The calculation of surface turbulent fluxes from synoptic information. (In Russian). *Meteor. Gidrol.* 8, 10-19.

Charnock, H., 1955: Wind stress on a water surface. *Quart. J. Roy. Meteor. Soc.*, 81, 639-642.

Clarke, R.H., 1970a: Observational studies in the atmospheric boundary layer. *Quart. J. Roy. Meteor. Soc.*, 96, 91-114.

Clarke, R.H., 1970b: Recommended methods for the treatment of the boundary layer in numerical models. *Austr. Meteor. Mag.*, 18, 51-73.

Clarke, R.H., 1971: Depth of the boundary layer. *Atmos. Environm.*, 5, 67-69.

Clarke, R.H., 1974: Attempts to simulate the diurnal course of meteorological variables in the boundary layer. (In Russian). *Izvestiya Acad. Sci. USSR*, 10, 600-612.

Clarke, R.H., 1975: Note on baroclinicity and the inverse behaviour of surface stress and wind turning in the boundary layer. *Beitr. Phys. Atm.*, 48, 46-50.

Clarke, R.H., A.J. Dyer, R.R. Brook, G.D. Reid and A.J. Troup, 1971: The Wangara experiment: boundary layer data. Commonw. Sci. Ind. Res. Org. Austr., Div. Meteor. Phys. Techn., Paper 19, 1-21.

Clarke, R.H. and G.D. Hess, 1973: On the appropriate scaling for velocity and temperature in the planetary boundary layer. *J. Atmos. Sci.*, 30, 1346-1353.

Clarke, R.H. and G.D. Hess, 1974: Geostrophic departure and the functions A and B of Rossby-number similarity theory. *Boundary-Layer Meteor.*, 7, 267-287.

Clarke, R.H. and G.D. Hess, 1975: On the relation between surface wind and pressure gradient, especially in lower latitudes. *Boundary-Layer Meteor.*, 9, 325-339.

Counihan, J., 1975: Adiabatic atmospheric boundary layers: a review and analysis of data from the period 1880-1972. *Atmosph. Environm.*, 9, 871-905.

Csanady, G.T., 1967: on the "resistance law" of a turbulent Ekman layer. *J. Atmosph. Sci.*, 24, 467-471.

Csanady, G.T., 1972: Geostrophic drag, heat and mass transfer coefficient for for diabatic Ekman layer. *J. Atmos. Sci.*, 29, 488-496.

Deacon, E.L., 1973: Geostrophic drag coefficients. *Boundary-Layer Meteor.*, 5, 321-340.

Deardorff, J.W., 1970a: Preliminary results from numerical integrations of the unstable planetary boundary layer. *J. Atmosph. Sci.*, 27, 1209-1211.

Deardorff, J.W., 1972a: Parameterization of the planetary boundary layer for use in general circulation models. *Mon. Weather Rev.*, 100, 93-106.

Deardorff, J.W., 1972b: Numerical investigation of neutral and unstable planetary boundary layers. *J. Atmosph. Sci.*, 29, 91-115.

Deardorff, J.W., 1973: An explanation of anomalously large Reynolds stresses within the convective planetary boundary layer. *J. Atmosph. Sci.*, 30, 1070-1076.

Deardorff, J.W., 1974a: Three-dimensional numerical study of the height and mean structure of a heated planetary boundary layer. *Boundary-Layer Meteor.*, 7, 81-106.

Deardorff, J.W., 1974b: Three-dimensional numerical study of turbulence in an extraining mixed layer. *Boundary-Layer Meteor.*, 7, 199-226.

Deardorff, J.W., G.E. Willis and D.K. Lilly, 1969: Laboratory investigation of non-steady penetrative convection. *J. Fluid Mech.*, 35, 7-31.

Delsol, F., K. Miyakoda and R.H. Clarke, 1971: Parameterized processes in the surface boundary layer of an atmospheric circulation model. *Quart. J. Roy. Meteor Soc.*, 97, 181-208.

de Moor, G., 1975: La turbulence dans la couche limite atmospherique, 1ère partie, parameterisation des flux turbulents. *La Météorologie*, 6, 107-124.

de Moor, G. and J.C. Andre, 1975: La turbulence dans la couche limite atmospherique, 2ème partie, modelisation de la couche limite. *La Météorologie* 6, 179-195.

Djolov, G.D., 1974: Modeling of interdependent diurnal variation of meteorological elements in the boundary layer. *Khidrologiya i Meteorologiya* 23, 6, 3-20.

Djolov, G. and E. Syrakov, 1975: Formula for the mixing length in the planetary boundary layer. *Khidrologiya i Meteorologiya* 24, 6, 7-13.

Dubov, A.S., 1975: On parameterization of processes in the planetary boundary layer in the numerical weather prediction. (In Russian). *Trudy of the Main Geophys. Obs. (GGO)*, 362, 41-49.

Dubov, A.S. and N.V. Kucherov, 1972: On the question of the determination of heat and momentum fluxes over sea. (In Russian). *Trudy of the Main Geophys. Obs. (GGO)*, 282, 175-186.

Dubov, A.S. and L.R. Orlenko, 1970: On taking into account of processes in the boundary layer in the numerical modeling of the general circulation of the atmosphere. (In Russian). *Trudy of the Main Geophys. Obs. GGO)*, 256, 45-56.

Ekman, V.W., 1905: On the influence of the earth's rotation on ocean-currents. *Arkiv Mat., Astron. Fys. 2*, 11, 1-52.

Ekman, V.W., 1906: Beiträge zur Theorie der Meeresströmungen. *Ann. Meteor.*, 34, 423-430, 472-484, 527-540, 566-583.

Ertel, H., 1932: Allgemeine Theorie der Turbulenzreibung und des "Austausches". *Sitz. Ber. Preuss. Akad. Wiss. Berlin, phys.-math. Kl.* 26, 436-445.

Ertel, H., 1937: Tensorielle Theorie der Turbulenz. *Ann. Hydr.*, 65, 193-205.

Estoque, M.A., 1973: Numerical modeling of the planetary boundary layer. In: Workshop on Micrometeorology, *Amer. Meteor. Soc.*, 217-270.

Etling, D. and F. Wippermann, 1975: The height of the planetary boundary layer and of the surface layer. *Beitr. Phys. Atm.*, 48, 250-254.

Everson, P.J. and D.R. Davies, 1972: Ekman boundary layer interactions in a numerical model of the general circulation. *Quart. J. Roy. Meteor. Soc.*, 98, 412-419.

Exner, F.M., 1912: Zur Kenntnis der untersten Winde über Land und Wasser und der durch sie erzeugten Meeresströmungen. *Ann. Hydr.*, 40, 226-239.

Fiedler, F., 1972: The effect of baroclinity on the resistance law in a diabatic Ekman layer. *Beitr. Phys. Atm.*, 45, 164-173.

Fiedler, F. and H.A. Panofsky, 1972: The geostrophic drag coefficient and the "effective" roughness length. *Quart. J. Roy. Meteor. Soc.*, 98, 213-220.

Fischer, G., E. Heise and V. Renner, 1973: Der Einfluß der Bodenreibung auf die Entwicklung von Zyklonenwellen in einem numerischen Modell. *Beitr. Phys. Atm.*, 46, 157-181.

GARP Working Group on Numerical Experimentation 1974: The GARP programme on numerical experimentation. Rep. No. 5, 20 pp.

Gill, A.E., 1968: Similarity theory and geostrophic adjustment. *Quart. J. Roy. Meteor. Soc.*, 94, 586-588.

Gossard, E.E. and W.R. Moninger, 1975: The influence of a capping inversion on the dynamic and convective instability of a boundary layer model with shear. *J. Atmos. Sci.*, 32, 2111-2124.

Hanna, S.R., 1969: The thickness of the planetary boundary layer. *Atmosph. Environm.*, 3, 519-536.

Hasse, L. and M. Dunckel, 1974: Direct determination of geostrophic drag coefficients at sea. *Boundary-Layer Meteor.*, 7, 323-329.

Heise, E., 1975: Ein Vergleich verschiedener Parameterisationen der turbulenten vertikalen Flüsse von Impuls und sensibler Wärme an der Meeresoberfläche. *Hamburger Geophys. Einzelschr.*, 27, 121 S.

Hess, G.D., 1973: On Rossby-number similarity theory for a baroclinic planetary boundary layer. *J. Atmosph. Sci.*, 30, 1722-1723.

Hoxit, L.R., 1974: Planetary boundary layer wind in baroclinic conditions. *J. Atmosph. Sci.*, 31, 1003-1020.

Hoxit, L.R., 1975: Diurnal variations in planetary boundary-layer winds over land. *Boundary-Layer Meteor.*, 8, 21-38.

Kaimal, J.C., J.C. Wyngaard, D.A. Haugen, O.R. Coté, Y. Izumi, S.J. Caughey and C.J. Readings, 1975: Turbulence structure in the convective boundary layer. *J. Atmos. Sci.*, 33, 2152-2169.

v. Kármán, Th., 1930: Mechanische Ähnlichkeit und Turbulenz. *Nachr. Ges. Wiss. Göttingen, Math.-Phys. Kl*, 56-76.

Kazanskii, A.B. and A.S. Monin, 1960: On the turbulent regime above the surface layer of the atmosphere. (In Russian). *Izvestiya Acad. Sci., USSR, Geophys. Ser.*, 165-168.

Kazanskii, A.B. and A.S. Monin, 1961: On the dybamical interaction between the atmosphere and the earth's surface. (In Russian). *Izvestiya Acad. Sci., USSR, Geophys. Ser.*, 786-788.

Kitaigorodsky, S.A., 1968: On the calculation of aerodynamic roughness of the sea surface. *Izv. Akad. Sci., USSR., Atmos. Oceanic Phys.*, 4, 870.

Kitaigorodsky, S.A., 1970: The Physics of the Interaction between the Atmosphere and the Ocean. (In Russian). Gidrometeoizdat, Leningrad.

Kukharets, V.P. and L.R. Tsvang, 1976: On parameterization of turbulent heat flux in unstable boundary layer of the atmosphere. (In Russian). *Izvestiya Acad. Sci., USSR, Atmosph. Ocean. Phys.*, 12, 13-21.

Kurosaki, A., 1970: A theoretical study on the structure of the Ekman boundary layer (I). *Papers Meteor. Geophys.*, 21, 89-112.

Laikhtman, D.L., 1970: Physics of the boundary layer of the atmosphere. (In Russian). 2nd ed., Hydromet. Press, Leningrad, 341 pp.

Laikhtman, D.L. and A.M. Popov, 1976: The effects of thermal stratification of the atmospheric boundary layer on its integral characteristics. (In Russian). *Izvestiya Acad. Sci., USSR, Atmosph. Ocean. Phys.*, 12, 663-665.

La Valle, L. and P. Di Girolamo, 1975: Determination of the geostrophic drag coefficient. *Tellus* 27, 87-92.

Lettau, H., 1962: Theoretical wind spirals in the boundary layer of a barotropic atmosphere. *Beitr. Phys. Atm.*, 35, 195-212.

Lettau, H. and B. Davidson, 1957: Exploring the atmosphere's first mile. Pergamon Press, London, New York, Paris, Vol. 1, 578 pp., Vol. 2, 376 pp.

Leykauf, H., 1974: Experimentelle Bestimmung universeller Profile und deren Abhängigkeit von Parametern der atmosphärischen Grenzschicht. *Techn. Hochschule, Darmstadt D* 17, 222 S.

Lilly, D.K., 1968: Models of cloud-topped mixed layers under a strong inversion. *Quart. J. Roy. Meteor. Soc.*, 94, 292-309.

Lo, A.K., 1977: Boundary layer flow over gentle curvilinear topography with a sudden change in surface roughness. *Quart. J. Roy. Meteor. Soc.*, 103, 199-209.

Lykosov, V.N., 1972: A nonstationary problem on a planetary boundary layer of the atmosphere of the earth. (In Russian). *Izvestiya Acad. Sci., USSR, Atmosph. Ocean. Phys.*, 8, 142-155.

Lykosov, V.N. and L.N. Gutman, 1972: Turbulent boundary layer over a sloping underlying surface. (In Russian). *Izvestiya Acad. Sci., USSR, Atmosph. Ocean. Phys.*, 8, 799-809.

MacKay, K.P., 1971: Steady state hodographs in a baroclinic boundary layer. *Boundary-Layer Meteor.*, 2, 161-168.

Mahrt, L.J. and W. Schwerdtfeger, 1970: Ekman spirals for exponential thermal wind. *Boundary-Layer Meteor.*, 1, 137-145.

Melgarejo, J.W. and J.W. Deardorff, 1974a: Stability functions for the boundary-layer resistance laws based upon observed boundary-layer heights. *J. Atmosph. Sci.*, 31, 1324-1333.

Melgarejo, J.W. and J.W. Deardorff, 1974b: Stability functions for the boundary-layer resistance laws based upon observed boundary-layer heights. Inst. Meteor. Univ. Stockholm, Intern. Meteor. Inst. Stockholm, Rep. DM-11, 34 pp.

Mildner, P., 1932: Über die Reibung in einer speziellen Luftmasse in den untersten Schichten der Atmosphäre. *Beitr. Phys. fr. Atmosph.*, 19, 151-158.

Monin, A.S., 1950: Dynamical turbulence in the atmosphere. (In Russian). *Izvestiya Acad. Sci., USSR, Geogr. Geophys. Ser.* 14, 232-254.

Monin, A.S., 1970: The atmospheric boundary layer. *Annual Rev. Fluid Mech.*, 2, 225-250.

Monin, A.S. and A.M. Yaglom, 1971: Statistical fluid mechanics: mechanics of turbulence, Vol. 1, MIT Press, 769 pp.

Monin, A.S. and S.S. Zilitinkevich, 1967: Planetary boundary layer and large-scale atmospheric dynamics. Report of the Study Conference held at Stockholm on the Global Atmospheric Research Programme (GARP), Appendix V, 37 pp.

Monin, A.S. and S.S. Zilitinkevich, 1974: Similarity theory and resistance laws for a planetary boundary layer. *Boundary-Layer Meteor.*, 7, 391-397.

Ordanovich, A.E. and V.N. Ivanov, 1974: On parameterization of turbulent exchange in Ekman boundary layer. (In Russian). *Meteorologiya i Gidrologiya*, 6, 26-32.

Orlenko, L.R. and O.B. Shklyarevich, 1975: Peculiarities of the structure of the instationary horizontally homogeneous boundary layer of the atmosphere. (In Russian). *Trudy of the Main Geophys. Obs. (GGO)*, 362, 96-106.

Priestley, C.H.B., 1967: Handover in scale of the fluxes of momentum, heat, etc. in the atmospheric boundary layer. *Phys. Fluids, Suppl.*, 38-46.

Rao, K.S., J.C. Wyngaard and O.R. Coté, 1974: The structure of the two-dimensional internal boundary layer over a sudden change of surface roughness. *J. Atmos. Sci.*, 31, 738-746.

Rossby, C.G., 1932: A generalization of the theory of the mixing length with applications to atmospheric and oceanic turbulence. Meteor. Pap. Mass. Inst. Techn., 1, 5-36.

Rossby, C.G. and R.B. Montgomery, 1935: The layer of frictional influence in wind and ocean. Pap. Phys. Ocean. Meteor. Mass. Inst. Techn., 3, No. 3, 1-101.

Rüffer, A., 1976: Zur Bestimmung geostrophischer Spannungskoeffizienten aus Windprofilen. *Z. Meteor.*, 26, 339-345.

Ruzin, M.I., N.A. Boldyreva and T.A. Savelieva, 1963: Some results of the calculation of the coefficient of turbulent exchange in the boundary layer of the atmosphere. (In Russian). *Trudy Hydromet. Inst. Leningrad (LGMI)*, 15, 66-80.

Schneider-Carius, K., 1953: Die Grundschicht der Troposphäre. *Probl. kosm. Phys.*, 26, *Akad. Verlagsges.*, *Leipzig*, 168 S.

Shir, C.C., 1972: A numerical computation of airflow over a sudden change in surface roughness. *J. Atmos. Sci.*, 29, 304-310.

Soloviev, G.N., 1973: The determination of certain universal relationships in the atmospheric boundary layer. (In Russian). *Meteorologiya i Gidrologiya* 12, 59-64.

Speranskii, L.S., V.N. Lykosov and L.N. Gutman, 1975: Turbulent boundary layer over the curvilinear underlying surface in the presence of the horizontal warm advection. (In Russian). *Izvestiya Acad. Sci. USSR, Atmosph. Ocean. Phys.* 11, 435-445.

Stull, R.B., 1973: Inversion rise model based on penetrative convection. *J. Atmos. Sci.*, 30, 1092-1099.

Sundatarajan, A., 1975: Significance of the neutral height scale for the convective barotropic planetary boundary layer. *J. Atmos. Sci.*, 32, 2285-2287.

Swinbank, W.C., 1974: The geostrophic drag coefficient. *Boundary-Layer Meteor.*, 7, 125-127.

Taylor, G.I., 1915: Eddy motion in the atmosphere. *Phil. Trans. Roy. Soc. A* 215, 1-26.

Taylor, P.A., 1969: The planetary boundary layer above a change in surface roughness. *J. Atmosph. Sci.*, 26, 432-440.

Taylor, P.A., 1977a: Some numerical studies of surface boundary-layer flow above gentle topography. *Boundary-Layer Meteor.*, 11, 439-466.

Taylor, P.A., 1977b: Numerical studies of neutrally stratified planetary boundary layer flow above gentle topography. I: Two-dimensional cases. *Boundary-Layer Meteor.*, 12, 37-60.

Tennekes, H., 1974: The atmospheric boundary layer. *Phys. Today*, 27.

Tennekes, H., 1973: A model for the dynamics of the inversion above a convective boundary layer. *J. Atmosph. Sci.*, 30, 558-567.

Vager, B.G. and E.D. Nadezhina, 1973: The change of the planetary boundary layer regime under influence of a change of moisture of the surface. (In Russian). *Trudy Main Geophys. Obs. (GGO)*, 297, 165-172.

Vager, B.G. and E.D. Nadezhina, 1975a: The structure of the planetary boundary layer of the atmosphere in conditions of the horizontal inhomogeneity. (In Russian). *Izvestiya Acad. Sci., USSR, Atmosph. Ocean. Phys.*, 11, 565-573.

Vager, B.G. and E.D. Nadezhina, 1975b: Numerical experiments on the calculation of the horizontally inhomogeneous planetary boundary layer. (In Russian). *Trudy Main Geophys. Obs. (GGO)*, 326, 11-19.

Vager, B.G. and S.S. Zilitinkevich, 1968: Theoretical model of diurnal variations of meteorological fields. (In Russian). *Meteorologiya i Gidrologiya*, 7, 3-18.

Venkatesh, S. and G.T. Csanady, 1974: A baroclinic planetary boundary-layer model and its application to the Wangara data. *Boundary-Layer Meteor.*, 5, 459-473.

Voltsinger, N.E., D.L. Laikhtman and E.P. Rudenko, 1973: Influence of the baroclinicity on the planetary boundary layer structure. (In Russian). *Izvestiya Acad. Sci., USSR.*, 9, 919-927.

Vorob'eva, A.G., G.P. Kurbatkin and D.L. Laikhtman, 1968: The influence of a horizontally inhomogeneous boundary layer on the dynamics of large-scale processes. (In Russian). *Izvestiya Acad. Sci., USSR, Atmosph. Ocean. Phys.*, 4, 125-140.

Wieringa, J., 1974: Comparison of three methods for determining strong wing stress over Lake Flevo. *Boundary-Layer Meteor.*, 7, 3-19.

Wippermann, F., 1970: The two constants in the resistance law for a neutral barotropic boundary layer of the atmosphere. *Beitr. Phys. Atm.*, 43, 133-140.

Wippermann, F., 1971: A mixing-length hypothesis for the planetary boundary layer flow in the atmosphere. *Beitr. Phys. Atm.*, 44, 215-216.

Wippermann, F., 1972a: A note on the parameterization of the large-scale wind stress at the sea-surface. *Beitr. Phys. Atm.*, 45, 260-266.

Wippermann, F., 1972b: Empirical formulae for the universal functions $M_m(\mu)$ and N (μ) in the resistance law for a barotropic and diabatic planetary boundary layer. *Beitr. Phys. Atm.*, 45, 305-311.

Wippermann, F., 1972c: Universal profiles in the barotropic planetary boundary layer. *Beitr. Phys. Atm.*, 45, 148-163.

Wippermann, F., 1972d: Baroclinic effects on the resistance law for the planetary boundary layer of the atmosphere. *Beitr. Phys., Atm.*, 45, 244-259.

Wippermann, F., 1973a: The planetary boundary layer of the atmosphere. *Ann. Meteor. (Neue Folge)*, 7, 346 pp.

Wippermann, F., 1973b: Numerical study on the effects controlling the low-level jet. *Beitr. Phys. Atm.*, 46, 137-154.

Wippermann, F., 1974: The parameterization of the large-scale wind stress at the sea-surface (second note). *Beitr. Phys. Atm.*, 47, 227-234.

Wippermann, F., 1975: Properties of the thermal boundary layer of the atmosphere obtained with a pbl-model. *Beitr. Phys. Atm.*, 1, 30-45.

Wippermann, F., D. Etling and H. Leykauf, 1973: The effects of nonstationarity on the planetary boundary layer. *Beitr. Phys. Atm.*, 46, 34-56.

Wyngaard, J.C., S.P.S. Arya and O.R. Coté, 1974a: Some aspects of the structure of convective planetary boundary layer. *J. Atmosph. Sci.*, 31, 747-754.

Wyngaard, J.C., O.R. Coté and K.S. Rao, 1974b: Modeling of the atmospheric boundary layer. *Adv. Geophys.* 18A, 193-212.

Yamamoto, G., N. Yasuda and A. Shimanuki, 1968: Effect of thermal stratification on the Ekman layer. *J. Meteor. Soc. Jap.*, 46, 442-455.

Yordanov, D., 1974a: The influence of baroclinicity on the resistance law. (In Russian). *Izvestiya Acad. Sci., USSR, Atmosph. Ocean. Phys.*, 10, 647-649.

Yordanov, D., 1974b: Parameterization of baroclinicity influence in the planetary boundary layer. (In Russian). *Izvestiya Acad. Sci., USSR, Atmosph. Ocean. Phys.*, 10, 782-785.

Yordanov, D., 1974c: Determination of universal functions in the planetary boundary layer using experimental data. (In Russian). *Izvestiya Acad. Sci., USSR, Atmosph. Ocean. Phys.*, 10, 1087-1089.

Yordanov, D., 1974d: Comments on "Rossby number similarity theory for a baroclinic planetary boundary layer". *J. Atmosph. Sci.*, 31, 1938.

Yordanov, D., 1974e: A simple two-layer model of the planetary boundary layer. *Arch. Meteor. Geophys. Biokl.*, A 23, 217-226.

Yordanov, D., 1975: Simplified baroclinic model of the planetary boundary layer. (In Russian). *Izvestiya Acad. Sci., USSR, Atmosph. Ocean. Phys.*, 11, 630-634.

Yordanov, D. and D. Dimitrov, 1974: Investigating the influence of baroclinicity in the planetary boundary layer by means of the model of Zilitinkevich, Tchalikov. (In Bulg.). *Khidrologiya i Meteorologiya*, 23, 34-41.

Yordanov, D. and F. Wippermann, 1972: The parameterization of the turbulent fluxes of momentum, heat and moisture at the ground in a baroclinic planetary boundary layer. *Beitr. Phys. Atm.*, 45, 58-65.

Zilitinkevich, S.S., 1967a: On the dynamic and thermal interaction between the atmosphere and the ocean. (In Russian). *Izvestiya Acad. Sci., USSR, Atmosph. Ocean. Phys.*, 3, 1069-1077.

Zilitinkevich, S.S., 1967b: Mathematical modelling of the interaction between the atmosphere and the ocean with regard to the processes in a viscous layer. 14. IUGG General Assembly Switzerland, September 25-October 7. IAMAP-Sympos. on Mechanism of Exchange Processes at the Earth-Atmosphere Interface, 18 pp.

Zilitinkevich, S.S., 1970: Dynamics of the atmospheric boundary layer. (In Russian). Hydromet. Press. Leningrad, 291 pp.

Zilitinkevich, S.S., 1972a: Asympotic formulae for the height of Ekman boundary layer. (In Russian). *Izvestiya Acad. Sci., USSR, Atmosph. Ocean. Phys.*, 8, 1086-1090.

Zilitinkevich, S.S., 1972b: On the determination of the height of the Ekman boundary layer. *Boundary-Layer Meteor.*, 3, 141-145.

Zilitinkevich, S.S., 1973: Shear convection. *Boundary-Layer Meteor.*, 3, 416-423.

Zilitinkevich, S.S., 1975: Resistance laws and prediction equations for the depth of the planetary boundary layer. *J. Atmosph. Sci.*, 32, 741-752.

Zilitinkevich, S.S. and D.V. Chalikov, 1968a: On the momentum, heat and moisture exchange in the process of interaction between in the atmosphere and underlying surface. (In Russian). *Izvestiya Acad. Sci., USSR, Atmosph. Ocean. Phys.*, 4, 765-772.

Zilitinkevich, S.S. and D.V. Chalikov, 1968b: On the vertical structure of the planetary boundary layer at unstable stratification. (In Russian). *Meteorologya i Gidrologiya*, 2, 11-26.

Zilitinkevich, S.S. and J.W. Deardorff, 1974: Similaritv theory for the planetary boundary layer of time-dependent height. *J. Atmosph. Sci.*, 31, 1449-1452.

Zilitinkevich, S.S. and D.L. Laikhtman, 1965: On the closure of the system of equations of turbulent motion for the boundary layer of the atmosphere. (In Russian). *Trudy Main Geophys. Obs. (GGO)*, 167, 44-48.

Zilitinkevich, S.S., D.L. Laikhtman and A.S. Monin, 1967: Dynamics of the atmospheric boundary layer. (In Russian). *Izvestiya Acad. Sci., USSR, Atmosph. Ocean. Phys.*, 3, 297-333.

Zilitinkevich, S.S. and A.S. Monin, 1974: The similarity theory for the boundary layer of the atmosphere. (In Russian). *Doklady Acad. Sci., USSR*, 216, 536-539.

CHAPTER 3 - TROPICAL BOUNDARY LAYER

3.1 Introduction

The tropical boundary layer is being considered separately because of its fundamental importance to the whole of the atmospheric motion system and because of its differences from boundary layers at mid and higher latitudes. The tropical regions cover about one half the earth's surface and are the direct recipients of most of the solar radiation. The oceanic tropical boundary layer provides "considerably more than half" (Malkus, 1962, p. 94) of the water evaporated into the atmosphere. This water vapour is a large potential energy source for convective and synoptic scale motions throughout the troposphere.

The tropical regions are generally considered to be contained within $\pm 30^{\circ}$ latitude. However, the distinction between a tropical and mid-latitude boundary layer is not precise. Many of the concepts presented in Chapters 1 and 2 also pertain to tropical boundary layers. One feature that changes as one moves towards the equator is the coriolis parameter, f. This affects the whole of tropical dynamics, including the boundary layer. Another important difference is the relatively high moisture content values of the tropical boundary layer over the oceans. Yet over some tropical land regions the air is exceedingly dry. It is not possible to define precisely what a tropical boundary layer is nor its geographical extent. There is a continuous transition between the tropics and mid-latitudes. The distinctions are affected by season, surface conditions (ocean or land) and other parameters. In this chapter the features of tropical boundary layers will be presented with the emphasis on the oceanic tropical boundary layer.

The tropical oceanic boundary layer has been the subject of a recent intensive experimental study; the GATE Boundary-Layer Sub-Programme (GATE Report No. 5, WMO-ICSU) as part of the GARP Atlantic Tropical Experiment (GATE). The results of the GATE should greatly increase our understanding of the tropical boundary layer. Unfortunately, at the time of preparation of this Technical Note only preliminary GATE results were available. These results were available in GATE Reports 14(I) and 14(II) and from the abstracts of the International Scientific Conference on the Energetics of the Tropical Atmosphere (Tashkent, U.S.S.R.; September 14-21, 1977), (available from WMO) and the WMO Technical Conference on GATE Data Utilization with Implication for Forecasting and Research in Tropical Countries (Cairo, Egypt; December 3-8, 1977), (available from WMO). An extensive (723 pages) and valuable source of GATE results is the Report of the U.S. GATE Central Program Workshop (published by NCAR).

In the preparation of this chapter considerable reference was made to Young et al., (Boundary Layer Group, Report of the Study Conference on the Development of Numerical Models for the Tropics, Exeter; April 4-10, 1976), Garstang and Betts (1974), and GATE Report No. 5.

3.2 Vertical Structure of the Tropical Boundary Layer

The tropical boundary layer is usually considered as being that layer between the surface and the trade-wind inversion. The trade-wind inversion is a very stably stratified layer that impedes the vertical transfers and provides a lid on cumulus convection. However, under certain conditions, the developing cumulus clouds will have sufficient energy to penetrate the inversion and this leads to a coupling and interchange between the surface layer and all layers up to the tropopause. In these cases the boundary-layer depth is either poorly or not defined. This latter condition, which is usually referred to as disturbed conditions, is very important for atmospheric dynamics. These disturbed conditions are very common in the vicinity of the intertropical convergence zone (ITCZ).

During those times when the trade-wind inversion effectively caps the boundary layer, called undisturbed conditions, the tropical boundary layer has a well defined vertical layering. These layers are characterized by (Riehl et al., 1951; Malkus, 1956):

(i) Surface Layer

A slight decrease in potential temperature with height and a stronger decrease in moisture content. This layer extends from the surface to a height of between 20 to 100 m.

(ii) Mixed Layer

Nearly constant potential temperature and constant or slightly decreasing moisture content. This layer extends from the top of the surface layer to about 100 m below cloud base.

(iii) Transition Layer

Stable potential temperature and a marked decrease in moisture content. This layer extends to cloud base and separates the regime of cloud convection and mechanical convection below.

(iv) Cloud Layer

Nearly a moist adiabatic temperature lapse rate and a decrease in moisture content. This layer extends from cloud base to the base of the inversion.

(v) Inversion Layer

Extremely stable temperature gradient and a strong decrease in humidity. This layer tops the undistrubed boundary layer.

The wind structure has a less pronounced stratification. Usually the wind increases from the surface to about 300-400 m and then decreases aloft due to the thermal wind effects. The wind direction slightly veers. In Figures 3.1 and 3.2 examples of

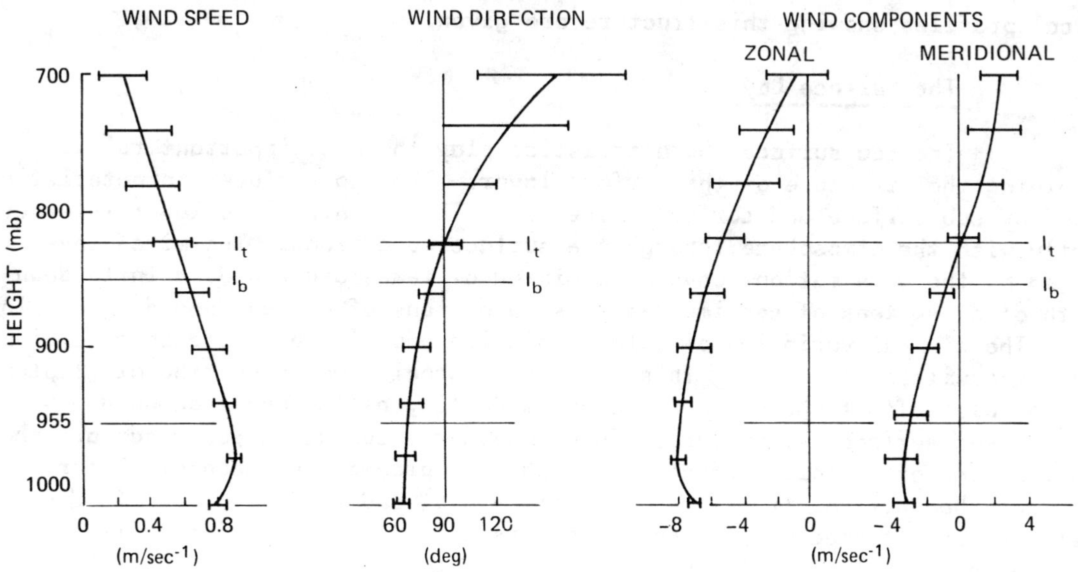

Figure 3.1 Average profiles of wind speed, direction and components. Horizontal bars indicate standard deviation of six days average in February 1969. I_t and I_b denote top and base of trade wind inversion. From Augstein et al., (1973).

Figure 3.2 Average profiles of temperature and specific humidity from three ships of the ATEX 800 km triangle. I_t and I_b denote top and base of trade wind inversion. From Augstein et al., (1973).

vertical profiles showing this tructure are given.

3.2.1 The Surface Layer

The sea surface characteristics play the most important role in determining the structure of the surface layer. The sea surface characteristics are defined by sub-surface and surface processes. The oceanic mixed layer interacts directly with the atmosphere through the surface conditions. The mixed layer is affected by the wave motions causing a mixing of temperature and salinity down to a depth of a few tens of centimetres to several tens of metres depending on wind speed. The diurnal variation of solar radiation causes the existence of a diurnal thermocline within the ocean. This and the internal waves occurring at greater depth may also affect the temperature and stability of the upper ocean and hence the sea surface temperature. The atmospheric surface layer winds are the primary source of oceanic surface waves and also provide the mechanical turbulent energy for the vertical transfers of sensible heat and water vapour. The sea surface temperature is affected by the heat transfers and at the same time influences them. The system of the atmospheric and oceanic boundary layers is a coupled, interacting system and it is not possible to consider them independently.

Another effect on the sea surface characteristic is precipitation. The precipitation may cool the surface water and will also change its salinity. The downdrafts associated with cumulus clouds may be reflected as colder surface patches, the so-called "cloud-induced footprint". The surface temperature will also be influenced to a considerable degree by the radiation balance. In summary there will be a balance of fluxes such that:

$$(1-\alpha)\ SR + LR^{\downarrow} - LR^{\uparrow} = H_o + LE_o + B_o + P_o \qquad (3.2.1)$$

- SR - solar radiation to the surface
- α - albedo, the ratio of reflected to incident solar radiation
- $LR^{\downarrow}$ - incoming long-wave radiation
- $LR^{\uparrow}$ - outgoing long-wave radiation
- H_o - sensible heat flux from the surface to the air
- E_o - evaporation from the surface
- L - latent heat of vaporization
- B_o - heat flux from the surface to the water
- P_o - heat loss due to precipitation

The terms LR, H_o, and B_o will all depend on the surface temperature. The albedo, α, H_o, LE_o and B_o will depend on the wind turbulence in the air and wind induced turbulence and waves in the water.

Each of these fluxes is potentially important and their determination is not simple. Generally the radiative fluxes are the easiest to measure but, at sea, it is necessary to carefully avoid the radiative effects of the ship. Polavarapu (1978) has demonstrated the value of using two specially shielded net radiometers, one on each side of the ship. He (1977) also deduced relationships between the components of the radiation budget for the tropical ocean. Kirillova and Egorov (1977) have investigated the influence of cloudiness on the net radiation.

The contribution due to precipitation has not been very well measured. There are indications that the cooling effect of the precipitation as well as the reduction in ocean surface-layer salinity may be important to the dynamics of the oceanic surface layer. The mesoscale downdrafts frequently associated with convective shower precipitation also leads to cooler surface water (Gautier, 1978). Over land the precipitation will also have an effect on the surface energy budget but its quantitative effect is not known.

The heat flux in the ocean may be determined from oceanographic measurements but is probably best determined by a residual or budget method. Over land, the corresponding flux of heat into the soil can be measured although its magnitude is usually small and usually sums to zero over 24 hours.

Boundary-layer meteorologists have been most concerned with the sensible and latent heat fluxes. Over the ocean the latent heat flux usually predominates, with values ranging from less than 50 to about 300 w/m^2 (Hoeber, 1975; Grose, 1975). Hoeber found that the variation at any one location could be 100% of the mean value with the variations most pronounced in the region of the ITCZ. The sensible heat flux is usually about an order of magnitude smaller and occasionally will be negative. The Bowen ratio, the ratio of sensible to latent heat fluxes, averages about 0.1 but shows considerable variability. Grose (1975), based on one ship's data, found values of between 0 and 0.3 with an average of 0.13. Elagina et al. (1975) measured a value of 0.19. Radikevich and Serjakov (1975), for a similar time period (GATE) but at the equator, found values from -0.05 to 0.2 with a mean of near 0.05. It is interesting to note that when the Bowen ratio is about 0.08 the turbulent energy production due to moisture fluctuations equals that due to temperature fluctuations.

The turbulent fluxes at the surface can be measured by the techniques described in Chapter 5 and Section 6.1. The Monin-Obukhov similarity theory (Chapter 1) should be equally valid over tropical land and water surfaces as over mid-latitude surfaces. Over tropical oceans it is necessary to include the effects on buoyancy of moisture in the definition of the Monin-Obukhov length. Preliminary results from GATE (e.g., Galushko et al., 1975) shows that the wind and humidity statistics are similar to earlier results as summarized in Busch (1973). Temperature turbulent statistics are still somewhat controversial and will be discussed below.

Several studies have been conducted to determine the bulk aerodynamic coefficients (see Section 6.1) over the ocean and it is generally accepted that $C_D \sim C_E \sim 1.3 \times 10^{-3}$. Based on preliminary GATE data values were found to be $C_D \sim C_E \sim 1.5 \times 10^{-3}$ (Wucknitz et al., 1975), $C_D = 1.1 \times 10^{-3}$, $C_E = 1.3 \times 10^{-3}$ (Muller-Glewe and Hinzpeter, 1975) and $C_D \sim 0.75 \times 10^{-3}$ for U (10 m) < 6 ms^{-1} and then increasing with wind speed to $C_D \sim 1.5 \times 10^{-3}$ for $U \sim 8$ ms $^{-1}$, $C_E = 0.75 \times 10^{-3}$

(Elagina et al., 1975). The Report of U.S. GATE Workship recommends a value of 1.4×10^{-3}. However, it is noted that the mean values do not entirely specify the fluxes so that the errors in using bulk formulae could be as large as a factor of two.

The fluctuations of temperature over the tropical ocean have been found to have different characteristics than over mid-latitude land or water surfaces. The temperature fluctuations have been shown (Phelps and Pond, 1971) to be dissimilar to the humidity fluctuations. For small scales ($\lambda < 30$ z) the temperature and humidity fluctuations are in-phase and highly correlated while for larger scales they tend to be out-of-phase but still correlated. This is likely to be related to the structure of the tropical boundary layer (Holland 1972). Galushko et al., (1975) found that the spectrum of temperature fluctuations during the GATE was different from over land results and generally agreed with the BOMEX results of Phelps and Pond (1971) for near neutral to moderate instabilities. For more unstable conditions ($-z/L > 0.25$) there was evidence that the temperature spectrum had more low frequency variance and was more similar to Phelps and Pond's San Diego (latitude 33°N) results than their BOMEX (latitude 15°N) results. Because of the dissimilarity of the temperature results the experimental determination of C_H have usually shown it to be different than C_D or C_E. Pond et al., (1971) found C_H to be very large and poorly determined because the sensible heat flux was not strongly dependent upon the air-sea temperature difference. To some extent, the temperature fluctuations and hence heat flux, were just carried along by the humidity fluctuations. Muller-Glewe and Hinzpeter (1975), Dunckel et al., (1974), Elagina et al., (1975) and Wucknitz et al., (1975) all found better correlation and consistently higher values for C_H (1.97, 2.1, 1.58, 1.8×10^{-3}, respectively). The question about the best value for C_H is still open. Recently, Schmidt et al., (1978) has shown that contamination of temperature probes by salt particles could lead to the observed differences between temperature and humidity fluctuations. These results cast serious doubts on our knowledge of temperature fluctuations at sea with relatively high humidity values.

Although reasonable estimates of the surface fluxes using the bulk aerodynamic method (except, perhaps, for heat flux) can be made during undisturbed conditions it is much more difficult to estimate the fluxes under disturbed conditions. The boundary layer is then expected to be non-steady and inhomogeneous. For example, organized cellular and vortex roll motions may form in unstable conditions (Lemone, 1973). Wucknitz et al., (1975) studied surface layer profiles before, during and after rain events and compared the deduced bulk aerodynamic coefficients with those deduced from undisturbed conditions. Surprisingly they found that, on the average, there was no significant difference in the deduced coefficients. Further results from GATE are being examined to further clarify this.

3.2.2 Mixed, Transitional and Cloud Layers

The mixed, transitional and cloud layers will be treated together because of their mutual interactions. The best described situations are for unstable stratifications over the oceans. These were studied during BOMEX (Holland et al., 1976), ATEX (Augstein et al., 1974), GATE and other experiments.

The unstable surface layer provides the buoyant energy to cause a well mixed, constant potential temperature layer above it. If there is large-scale subsidence and the buoyant forces are not too large then this mixed layer will be capped by the trade-wind inversion. This is essentially the situation described in Chapter 1. However, over the ocean there is frequently enough moisture in the rising air parcels that they reach their condensation level and form clouds at a height lower than the inversion base. The formation of the cloud layer, which is generally still capped by the trade-wind inversion, changes the structure of the upper part of the boundary layer. The temperature gradient instead of being dry adiabatic, is then nearly moist adiabatic.

Yanai et al., (1973) and Ogura and Cho, (1973) have shown that the upward mass flux within the clouds is larger than the mean upward mass flux and, hence, there must be compensating downward motion in the cloud free areas. The descending motion will have its maximum magnitude near cloud base. Because the descending air is warmed adiabatically this will result in the formation of a warm and hence stable layer at and below cloud base. This is called the transitional layer. The transitional layer is not expected to be found under clouds but, for undisturbed conditions, because the area of clouds is a small fraction of the total area the transitional layer should be evident in most locations.
Augstein et al. (1974) reported that more than 70% of all soundings obtained in the undisturbed trade-wind regime exhibited this form of layer structure.

Thus, there is an interaction between the buoyant turbulent energy produced at the surface which leads to the formation of clouds and the sinking motions which counteract it. Garstang (1973) has suggested that cumulus clouds or groups may be self-limiting through this interaction with the sub-cloud layer. The formation of the transitional layer will inhibit the further formation of clouds in the cloud-free areas. In addition the surface layer buoyancy is partially (perhaps 50%) due to moisture and as the clouds form the moisture is lost from the mixed layer. It is suggested that this gradually dries out the mixed layer. With the further development of clouds, precipitation and associated cloud downdrafts will also affect the sub-cloud layer and may significantly change the velocity fields and hence the low-level convergence (Ulanski et al., 1973).

If there is sufficient low level convergent or correspondingly large scale upward motion then the cloud ensembles above the mixed layer may develop to the extent that they penetrate the inversion layer and extend to the upper troposphere. In this case the top of the PBL cannot be identified. The heavy rain and intensive cloud circulation modify the air column down to the surface. These processes are strongly non-linear and not presently well understood. Arakawa and Shubert (1974) have developed a model to attempt to describe this case in large-scale models. Further insight into this problem should result from investigation of the GATE data.

If disturbed the mixed layer is suppressed by the descending motions and mesoscale effects. The depth of the mixed layer may be limited to near 300 m by an inversion. For GATE it was estimated that the mixed layer was affected by convective disturbances about 30% of the time. Within the mixed layer there are two important time scales; namely, the recovery time of the boundary layer, $h/C_i U$, where h is the boundary layer depth, C_i the bulk transfer coefficient for the transfer of property i, U the mean wind speed and a time scale determined by the divergence or convergence, D^{-1}. The ratio hD/C_iU can be considered as a characteristic parameter for the boundary layer.

Although most tropical boundary layers that have been studied were over the oceans the 1972 Venezuelan International Meteorological and Hydrological Experiment (VIMHEX) was conducted over land. The results of Dugan (1973) and Betts et al. (1974) have also shown this interaction between the cloud and sub-cloud layers. Dugan showed that the boundary layer becomes progressively more stable during the passage of a major convective storm.

The boundary layer over land at night or over colder surface waters, such as would occur in regions of coastal or equational upwelling, will have a stable stratification in its lowest levels. This type of boundary layer is not well understood. It seems likely that there would be little interaction between the surface layer and the upper layers and that the surface fluxes will be small.

3.3 Modelling the Tropical Boundary Layer

Much of the work on modelling the tropical boundary layer overlaps with modelling of the boundary layer in general (Chapter 4) so this section will be very short. Some work has been done on the use of turbulence models (Deardorff, 1974, 1976; Sommeria, 1976) to incorporate the effects of moisture (not precipitation), radiation and large-scale subsidence. Satisfactory results have been obtained for undisturbed tropical boundary layers but further work, particularly on the effects of the cloud layer, is needed. These models unfortunately require considerable computing time and their general use will be limited to a few centres.

One area of considerable progress over the last decade is the development of models of the dry convective boundary layer (Ball, 1960; Betts, 1973a; Deardorff, 1972; Tennekes, 1973; Carson, 1973; Stull, 1973, etc.). As described in Chapter 1 these models have been quite successful in dealing with the dry, convective boundary layer. Another type of modelling is the diagnostic studies of the boundary and cloud layers. Holland and Rasmusson (1973), Augstein et al. (1973), Nitta and Ebsensen (1974) and Brummer et al. (1974) have shown the vertical structure of the convective transports of heat, water and momentum through budget studies of the BOMEX and ATEX data. The stratification of the convective boundary layer was investigated by Aspliden (1971) and Betts (1973a,c). Recently models of marine stratocumulus convection by Schubert (1976), Deardorff (1976), Randall (1976) and Schubert et al. (1977) have extended the earlier work of Lilly (1968). These models predict the cloud heights, the convective fluxes and the paths of parcels.

Most of the above models are based on the thermodynamics and have neglected the wind dynamics. Geisler and Kraus (1969) have considered the ageostrophic mass flux and have shown with a simple model that this results in a deepening of the mixed layer. The sign and magnitude of the vertical shear of the horizontal wind may be critical in the maintenance of cumulus clouds and Garstang (1973) has suggested that if the profile has a maximum in the cloud layer then the convection increases and becomes more organized. Alternatively, if the maximum is in the sub-cloud layer, the convection dies. Further studies are still needed on the dynamical-thermodynamical interactions and effects.

Mesoscale numerical models have been developed for a range of convective problems. In addition to the work of Deardorff (1972) and Sommeria (1976) that has already been mentioned, there have been the studies of Hill (1974) and Pielke (1974). Hill used a two-dimensional simulation model of a cloud field over central Florida to show how the interaction between cumulus cells of different sizes and stages of development led to the suppression of some and the enhancement of others. The importance of the boundary-layer convergence in sea-breeze fronts in controlling the development and motion of thunderstorm complexes over a heated land surface was demonstrated by Pielke using a three-dimensional model. Anthes (1974a,b) has used numerical models to investigate the structure of hurricanes and mature tropical storms.

The inclusion of the mesoscale fluxes in large-scale models is still a serious problem. They may have to be treated in a probabilistic sense, in view of the basic intermittency of the mesoscale systems. This intermittency implies (a) uncertainty in mesoscale intensity and character, and (b) parameterization problems due to poor scale separation between the cumulus, meso-system, and large scales. The ensemble model of Arakawa and Schubert (1974), which predicts a spectrum of clouds and their thermodynamic transports from large-scale forcing functions, is an attempt at parameterization of the mesoscale processes. Other studies are referred to in Garstang and Betts (1974, p. 1201).

3.4 Mesoscale Boundary-Layer Processes

Satellite pictures have shown that even the undisturbed boundary layer has considerable mesoscale structure. The cumulus clouds tend to be in patches or lines of scales 100-1000 km rather than uniform sheets. The investigation of these "cloud clusters" was one of the principal objectives of the GATE. The distribution of clouds can be used to indicate the location of vertical advection of moisture from the mixed layer. As discussed previously (Section 3.2) the transport upward into the clouds exceeds the mean moisture supply from the surface and hence will dry out the sub-cloud layer. However, if there is well-organized mesoscale circulation the cloud system can import moist air from other locations and maintain its existence for a longer period of time. The clouds may form as open or closed cells (Krishnamurti, 1975) or lines (Kuettner, 1971). Each of these may play a significant role in the sub-cloud or cloud layer transport. Lemone and Pennell (1976) showed, for a particular fair weather case, that the sub-cloud-layer roll-vortex system accounted for most of the flux. These kilometre-scale transports probably accounted for the counter-gradient momentum flux in the upper two-thirds of the sub-cloud layer.

A strong relationship between clouds or cloud patches and sub-cloud layer fluxes has been documented using data from GATE, and earlier experiments. In the most suppressed conditions, the clouds are probably passive, forming from moist turbulent elements brought up in the upward moving air of a sub-cloud eddy. With more active moist convection, however, the sub-cloud layer will feel the direct effect of the convection overhead; cloud updrafts or "roots", will play a part in sub-cloud layer dynamics as soon as the cloud begins drawing air from the sub-cloud layer.

These "roots" and the compensating downward motions then begin enhancing the transfer of mass, heat, moisture and momentum across the cloud-sub-cloud layer interface. These transfers occur in a wide range of scales determined by the various cloud sizes and heights and cannot be represented by simple down-gradient formulae. In fact examples of vertical fluxes against the local gradients are common. The contribution of eddies up to the scale of the cloud patches has to be included but remains poorly known. Where in the hierachy of the cumulus enhancement these effects become more important is not yet known. More insight into the relationship of sub-cloud layer fluxes to the distribution and intensity of the cumulus clouds overhead will, it is hoped, be gained from GATE B-scale and C-scale data of Phase III.

3.5 Large-Scale Dynamics of the Tropical Boundary Layer

An important factor in tropical boundary layer dynamics is the diminishing importance of the Coriolis force as the equator is approached. Although the geostrophic approximation does hold to quite low latitudes the wind field and mass field are less strongly coupled. One result is that the accelerations due to temporal or advective effects became relatively more important thus leading to the possible existence of several types of flows that are uncommon at high latitudes. For example when the local acceleration exceeds the Coriolis acceleration the result is a Stokes-type flow.

If there are large temporal accelerations due to the presence of synoptic-scale wave oscillations a quasi-Stokes type flow may occur near the equator. The transition to quasi-Ekman flow at higher latitudes may produce enhanced vertical motions (pumping) near associated critical latitude (Holton et al., 1971). The critical latitude would be where the Coriolis frequency equals the wave frequency. The advective regime occurs at low latitudes with quasi-steady motions; good examples being the cross equational monsoonal flows off the west coast and east coasts of Africa. Majrt (1972) has studied such situations with a simple model.

There has been extensive research into discovering the mechanisms for the formation of the ITCZ and tropical cloud systems. Most of this research is beyond the scope of this Technical Note and the reader is referred to the reports of Krishnamurti (1976), Garstang and Betts (1974), p. 1201), and the U.S. GATE Central Program Workshop.

References

Anthes, R.A., 1974a: Data assimilations and initialization of hurricane prediction models. *J. Atmos. Sci.*, 31, 702-719.

Anthes, R.A., 1974b: The dynamics and energetics of mature tropical cyclones. *Rev. Geophys., Space Phys.*, 12, 495-522.

Arakawa, A. and W.H. Schubert, 1974: Interaction of a cumulus cloud ensemble with the large-scale environment; Part I. *J. Atmos. Sci.*, 31, 674-701.

Aspliden, C., 1971: On the energy distribution in the tropical atmosphere. Ph.D. Dissertation, Florida State University, 232 pp.

Augstein, E.H., R. Riehl, R. Ostapoff and V. Wagner, 1973: Mass and energy transports in an undisturbed Atlantic trade wind flow. *Mon. Wea. Rev.*, 101, 101-111.

Augstein, E.H., H. Schmidt and F. Ostapoff, 1974: The vertical structure of the atmospheric planetary boundary layer in undisturbed trade winds over the Atlantic Ocean. *Boundary-Layer Meteor.*, 6, 129-150.

Ball, F.K., 1960: Energy changes involves in disturbing a dry atmosphere. *Quart. J. Roy. Meteor. Soc.*, 86, 483-491.

Betts, A.K., 1973a: A non-precipitating cumulus convection and its parameterization. *Quart. J. Roy. Meteor. Soc.*, 99, 178-196.

Betts, A.K., 1973b: A composite mesoscale cumulonimbus budget. *J. Atmos. Sci.*, 30, 597-610.

Betts, A.K., 1973c: A relationship between stratification, cloud depth and permitted cloud radii. *J. Appl. Meteor.*, 12, 890-893.

Betts, A.K., F.J. Dungan and R.W. Grover, 1974: Residual errors of the VIZ radiosonde hygristor as deduced from observations of the sub-cloud layer structure. *Bull. Amer. Meteor. Soc.*, 55, 1123-1125.

Brummer, B., E. Augstein and H. Riehl, 1974: On the low level wind structure of the Atlantic trades. *Quart. J. Roy. Meteor. Soc.*, 100, 109-121.

Busch, N.E., 1973: On the mechanics of atmospheric turbulence. Workshop on Micrometeorology (Haugen, ed.), Amer. Meteor. Soc., Boston, 1-66.

Carson, D.J., 1973: The development of a dry inversion-capped convectively unstable boundary layer. *Quart. J. Roy. Meteor. Soc.*, *99*, 450-467.

Deardorff, J.W., 1972: Numerical investigation of neutral and unstable planetary boundary layer. *J. Atmos. Sci.*, *29*, 91-115.

Deardorff, J.W., 1974: Three-dimensional numerical study of turbulence in an entraining mixed layer. *Boundary-Layer Meteor.*, *7*, 199-226.

Deardorff, J.W., 1976: On the entrainment rate of a strato cumulus topped mixed layer. *Quart. J. Roy. Meteor. Soc.*, *102*, 563-582.

Dungan, F.J., 1973: The thermodynamic structure of the cumulus sub-cloud layer. Atmospheric Science Paper No. 205, Colorado State University, 97 pp.

Dunckel, M., L. Hasse, L. Krugermeyer, D. Schriever and J. Wucknitz, 1974: Turbulent fluxes of momentum, heat and water vapour in the atmospheric surface layer at sea during ATEX. *Boundary-Layer Meteor.*, *6*, 81-106.

Elagina, L.G., B.M. Koprov and Y.A. Volkov, 1975: Characteristics of the turbulent exchange between the ocean and atmosphere in the equatoral Atlantic. GATE Report 14(I), 260-266.

Galushko, U.V., U.N. Ivanov, I.V. Nekrasov, V.P. Pudov, A.V. Rostkov and A.A. Shushkov, 1975: Turbulent characteristic measurements in the marine boundary layer during GATE. GATE Report 14(II), 237-261.

Garstang, M., 1973: Proc. Mainz Symposium on Atmospheric Boundary Layer.

Garstang, M. and A.K. Betts, 1974: A review of the tropical boundary layer and cumulus convection; structure, parameterization and modeling. *Bull. Amer. Meteor. Soc.*, *55*, 1195-1205.

Gautier, C., 1978: Some evidence of cool surface water pools associated with mesoscale downdrafts during GATE. *J. Phys. Oceano.*, *8*, 162-166.

Geisler, J.E. and E.B. Kraus, 1969: The well mixed Ekman boundary layer. *Deep-Sea. Res.*, *16*, *Suppl.*, 73-84.

Grose, P.L., 1975: Preliminary estimates of turbulent energy fluxes in the Western B-Scale. GATE Report 14(I), 252-259.

Hill, G.E., 1974: Factors controlling the size and spacings of cumulus clouds as revealed by numerical experiments. *J. Atmos. Sci.*, *31*, 646-673.

Hoeber, H., 1975: Variations of evaporation in the B-scale area of GATE. GATE Report 14(I), 245-257.

Holland, J.Z., 1972: Comparative evaluation of some BOMEX measurements of sea surface evaporation, energy flux and stress. *J. Phys. Oceano.*, 2, 476-486.

Holland, J.Z. and E.M. Rasmusson, 1973: Measurements of the atmospheric mass, energy and momentum over a 500 kilometer square of tropical ocean. *Mon. Wea. Rev.*, 101, 41-55.

Holton, J.R., J.M. Wallace and J.A. Young, 1971: On boundary layer dynamics and the ICTZ. *J. Atmos. Sci.*, 28, 275-280.

Kirillova, T.V. and B.N. Egorov, 1977: The influence of cloudiness on global radiation. Paper presented at WMO Conference on the Energetics of Tropical Atmospheres, Tashkent, USSR.

Krishnamurti, R., 1975: On cellurlar cloud pattern. *J. Atmos. Sci.*, 32, 1354-1383.

Krishnamurti, T.N., (editor), 1976: Report of the Study Conference on the Development of Numerical Models for the Tropics, Exeter, 4-10 April, 1976: The GARP Programme on Numerical Experimentation, Report No. 13, (available from WMO).

Kuettner, J.P., 1971: Cloud bands in the earth's atmosphere-observations and theory. *Tellus*, 23, 404-426.

Lemone, M.A., 1973: The structure and dynamics of horizontal roll vortices in the planetary boundary layer. *J. Atmos. Sci.*, 30, 1077-1091.

Lemone, M.A. and W.T. Pennell, 1976: The relationship of trade wind cumulus distribution to sub-cloud layer structure and fluxes. *Mon. Wea. Rev.*, 104, 524-539.

Lilly, D.K., 1968: Models of cloud-topped mixed layers under a strong inversion. *Quart. J. Roy. Meteor. Soc.*, 94, 292-309.

Mahrt, L.J., 1972: A numerical study of the influence of advective accelerations in an idealized, low-latitude, planetary boundary layer. *J. Atmos. Sci.*, 29, 1477-1484.

Malkus, J.S., 1956: On the maintenance of the trade winds. *Tellus*, 8, 335-350.

Malkus, J.S., 1962: Large-scale interactions. *The Sea.* M.N. Hill (editor), 88-294.

Muller-Glewe, J. and H. Hinzpeter, 1975: Turbulent fluxes in the ITCZ during during GATE Phase III at Station 27. GATE Report 14(I), 224-232.

Nitta, T. and S. Esbensen, 1974: Heat and moister budget analysis using BOMEX data. *Mon. Wea. Rev.*, 102, 17-28.

Ogura, Y. and H.R. Cho, 1973: Diagnostic determination of cumulus cloud populations from observed large-scale variables. *J. Atmos. Sci.*, 30, 1276-1286.

Phelps, G.T. and S. Pond, 1971: Spectra of the temperature and humidity fluctuations and the fluxes of moisture and sensible heat in the marine boundary layer. *J. Atmos. Sco.*, 28, 918-928.

Pielke, R., 1974: A three-dimensional numerical model of the sea breezes over South Florida. *Mon. Wea. Rev.*, 102, 115-139.

Polavarapu, R.J., 1977: Estimation of net radiation from global solar radiation or total downward radiation measurements. Presented at Radiation Sub-Programme Meeting, Leningrad, November 21-26 (to be published in ATMOSPHERE-OCEAN).

Polavarapu, R.J., 1978: Measurement of net radiation from shipboard sensors. *J. Appl. Meteor.*, 1062-1067.

Pond, S., G.T. Phelps, J.E. Paquin, G.A. McBean and R.W. Stewart, 1971: Measurements of the turbulent fluxes of momentum, moisture and sensible heat over the ocean. *J. Atmos. Sci.*, 28, 901-917.

Radikevich, V.M. and E.I. Serjakov, 1975: The interaction between the surface and the atmosphere at the equator. GATE Report 14(I), 267-275.

Randall, D., 1976: The interaction of the planetary boundary layer with large-scale circulation. Ph.D. Thesis, Dept. of Meteor., Univ. of California, Los Angeles.

Riehl, H., T.C. Yeh, J.S. Malkus and N.E. LaSeur, 1951: The north-east trade of the Pacific Ocean. *Quart. J. Roy. Meteor. Soc.*, 77, 598-626.

Schmitt, K.F., C.A. Friehe and C.H. Gibson, 1978: Humidity sensitivity of atmospheric temperature sensors by salt contamination. *J. Phys. Oceano.*, 8, 151-161.

Schubert, W.H., 1976: Experiments with Lilly's cloud-topped mixed layer model. *J. Atmos. Sci.*, 33, 436-446.

Schubert, W.H., J. Wakefield, E. Steiner and S.K. Cox, 1977: Marine strato-cumulus convection. Atmospheric Science Rept., Colorado State University, Fort Collins, Colorado, (submitted to J. Atmos. Sci.).

Sommeria, G., 1976: Three-dimensional simulation of turbulent processes in an undisturbed trade-wind boundary layer. *J. Atmos. Sci.*, 33, 216-241.

Stull, R.B., 1973: Inversion rise model based on penetrative convection. *J. Atmos. Sci.*, 30, 1092-1099.

Tennekes, H., 1973: A model for the dynamics of the inversion above a convective boundary layer. *J. Atmos. Sci.*, 30, 558-576.

Ulanski, S.L., R.K. Hadlock and M. Garstang, 1973: The role of convection is surface property and velocity fluctuations. *Boundary-Layer Meteor.*, 6, 183-195.

WMO: GATE Boundary-Layer Sub-Proramme. GATE Report No. 5.

WMO: Preliminary Scientific Results of the GATE. GATE Reports 14(I) and 14(II).

Yanai, M.S., S. Eshensen and J. Chu, 1973: Determination of bulk properties of tropical cloud clusters from large-scale heat and moisture budgets. *J. Atmos. Sci.*, 30, 611-627.

Young, J., E. Augstein, D. Carson, M. Lemone and G. Sommeria, 1976: Boundary Layers. In the Report of the Study Conference on the Development of Numerical Models for the Tropics, Exeter, 4-10 april 1976. (See Krishnamurti, 1976).

CHAPTER 4 - NUMERICAL MODELLING OF THE ATMOSPHERIC BOUNDARY LAYER

4.1 Introduction

Processes in the planetary boundary layer (PBL) have come to play an important role in many areas of meteorology as well as in many applications like agriculture, air pollution studies, urban planning, etc. The increased knowledge about atmospheric turbulence has made it possible to physically model important aspects of the PBL. The Monin-Obukhov similarity theory has made it possible to describe the surface layer quite well. For the Ekman layer, extending from the top of the surface layer to the free atmosphere, no such convenient theory exists. Non-stationary, non-homogeneous conditions play important roles in shaping the structure of the Ekman layer. Radiation and condensation enter the boundary layer problems in a complicated way. In many applications we meet situations that are far from the idealized conditions of the laboratory and simplified theories, but many of these processes might still be incorporated if we make use of approximate numerical models.

Many numerical models have already been developed for a wide range of applications with different degrees of sophistication. The demands on the models vary as well as the formulation, or parameterization, of basic physical processes. Computer capacity restrictions very often determine the structure of a specific model.

Numerical models of the atmospheric boundary layer are today capable of simulating, or coping with, a number of different aspects of atmospheric motions. Very sophisticated models are used in testing new hypotheses about the PBL structure. Others try to deal with air pollution dispersion and diffusion, for prediction, local weather forecasting, runway ice, etc. It is natural to divide our discussion into two kinds of models:

(a) Research models for studies in atmospheric turbulence and the structure and physics of the PBL.

(b) Operational models for various applications.

Naturally there is a feedback between the two classes.

The main difference in approach between the two kind of models can be illustrated by the following table showing the CPU-time needed for running a 24 hour forecast of PBL-simulation with different numerical PBL-models:

(a)	Deardorff's model (1974) 3-d model, 64,000 gridpoints, Δt = 6 seconds, CDC-7600.	350h/24h forecast

(b) Pielke's sea breeze model (1973) — 6h/24h forecast
No "physics". 3-d CDC-6600.

(c) Tapp and White's sea breeze model (1976) — 52 minutes/24h
3-d. No physics. IBM 360/195.

(d) Long and Shaffer's model (1975) — 10-15 minutes/24h
3-d. "All" physics included.
IBM 360/195.

(e) One-dimensional models (Yu, Yamada and Mellor, Bodin, Gutman etc.). All physics included. — ~1/2 minute/24h
IBM 360/195.

It is obvious that only the last two kind of models can meet operational requirements at the present time. The discussion in this Chapter will concentrate on the latter kind of models, i.e., model formulations suitable for various applications.

4.2 Turbulence Closure and Higher Order Modelling

In most definitions of the boundary layer turbulence is the central feature. The turbulent exchange is the main factor shaping the structure of the boundary layer in terms of the mean quantities Θ, potential temperature; Q, specific humidity; and U and V the westerly and southerly components of the wind. Considerable efforts have been undertaken by PBL modellers to come up with effective schemes for representing effectively the turbulent exchange of the momentum and heat in the equations of motion. As is well known, and discussed in the first chapter, the momentum and heat fluxes arise in the equations of motion (for the mean motion) as Reynold's terms, i.e. the correlations

$$\overline{u_i u_j},\ \overline{\theta_{ui}} \text{ and moisture flux } \overline{q u_i} \qquad (4.2.1)$$

j, i = 1, 2, 3, where it is assumed that the wind field, temperature field and moisture field can be written as

$$\text{wind} = U_i + u_i$$

$$\text{potential temperature} = \Theta + \theta \qquad (4.2.2)$$

$$\text{specific humidity} = Q + q$$

where upper case quantities represent averages and lower case quantities turbulent deviations from the average with the property

$$\overline{u_i} = \overline{\theta} = \overline{q} \equiv 0. \qquad (4.2.3)$$

The closure problem then boils down to expressing the correlations (4.2.1) as some functions of external parameters and internal, mean, parameters.

4.2.1 K-theory

The classical approach is to assume a formal analogy with molecular heat conduction and putting, if we assume approximate horizontal homogeneity and stationarity

$$- \overline{uw} = K_M \frac{\partial U}{\partial z}$$

$$- \overline{vw} = K_M \frac{\partial V}{\partial z} \qquad (4.2.4)$$

$$- \overline{\theta w} = K_H \frac{\partial \Theta}{\partial z}$$

$$- \overline{qw} = K_E \frac{\partial Q}{\partial z}$$

where K_M, K_H and K_E are eddy diffusivities (eddy exchange or eddy viscosity coefficients).

Prandtl (1925) introduced a mixing length, ℓ, and assumed that

$$K_M = \ell^2 \cdot \left|\frac{\partial \overline{V}}{\partial z}\right| = \ell^2 S \qquad (4.2.5)$$

where,

$$S = \left\{ \left(\frac{\partial U}{\partial z}\right)^2 + \left(\frac{\partial V}{\partial z}\right)^2 \right\}^{1/2}$$

for neutral stratification.

Different K-formulations have been used extensively in various PBL-models. A summary of earlier assumptions about K is discussed in Chapter 2. The simple form (4.2.5) suffers from many deficiencies and has been modified and generalized in different ways since 1925. Appendix I summarizes some K-formulations used by different modellers. Different K-formulations are discussed in Estoque (1963), Shir and Bornstein (1977), Yu (1977) and Bodin (1977). Yu's paper is of particular interest since it compares different formulations within the framework of the same numerical model. Clarke (1974) also discusses comparisons between different formulations applied on the averaged Wangara data (Clarke et al., 1971).

Limitations of K-theory:

Most K-theories are incapable of describing properly the convective boundary layer under strong instability. Neither can they describe flows over a non-homogeneous surface unless the scale is fairly large. The non-stationary transitions between different regimes, for example between night and day, are badly handled. K-theory also always predicts "down-the-gradient" transfer, which seems not to be the case at all times in the boundary layer under disturbed conditions or strongly unstable stratification. To be able to model properly processes like this it seems necessary to turn to more complicated parameterizations or patch together different approaches. Both of these ideas have been used in PBL-modelling.

4.2.2 The Surface Layer

Close to the ground, where the atmosphere responds quickly to variations in surface properties, the assumptions of stationarity and horizontal homogeneity are fairly good. In the surface layer we can make use of Monin-Obukhov's similarity theory, (see Chapter 1) which makes it possible to eliminate explicit numerical calculations in the lowest 50 meters of the PBL. The Monin-Obukhov theory is stated in the form of a dimensionless wind shear and lapse rate.

$$\frac{kz}{u_*}\frac{\partial U}{\partial z} = \phi_M(z/L) \qquad (4.2.6)$$

$$\frac{kz}{\theta_*}\frac{\partial \Theta}{\partial z} = \phi_H(z/L) \qquad (4.2.7)$$

where k is von Karman's constant (0.35 - 0.4), $u_* = \sqrt{\frac{\tau}{\rho}}$, $\Theta_* = -\ \overline{w\Theta}/u_*$, L = Monin-Obukhov's length = $- u^3_*/\frac{g}{\theta} k\ \overline{w\theta}$ and the surface stress, $\tau = -\rho\ \overline{uw}$. Φ_M and ϕ_H are the non-dimensional, universal, functions, of the stability parameter z/L which have to be determined from observations. The most accepted forms of Φ_M and Φ_H are due to Businger et al. (1971) (see Section 6.1 of this Note). The functions Φ_M and Φ_H given by Businger are easy to integrate giving U = U(z), V = V(z) and $\theta = \theta(z)$ for $z < z_s$, z_s being the top of the surface layer. Barker and Baxter (1975) have suggested an approximate way of integrating (4.2.6) and (4.2.7). This is also discussed in Paulson (1970) and Nickerson and Smiley (1975).

If we want the K's for the surface layer these are given by

$$K_M = ku_* z/\phi_M = \frac{k^2 z^2}{\phi^2_M}\cdot\frac{\partial U}{\partial z}$$

$$K_H = ku_* z/\phi_H = \frac{k^2 z^2}{\phi_H \cdot \phi_M}\cdot\frac{\partial U}{\partial z} \qquad (4.2.8)$$

by using

$$u_* = \sqrt{\frac{\tau}{\rho}} = \left(K_M \frac{\partial U}{\partial z}\right)^{1/2}$$

Equations (4.2.8) imply a mixing length in the surface layer,

$$\ell_M = \frac{kz}{\phi_M}$$

where ϕ_M is stability dependent.

By means of surface layer theory it is then possible to split up the numerical model into two parts as shown in Figure 4.1.

Above the surface layer we can make use of different formulations of turbulent exchange provided we make a proper matching of the surface layer and the Ekman layer.

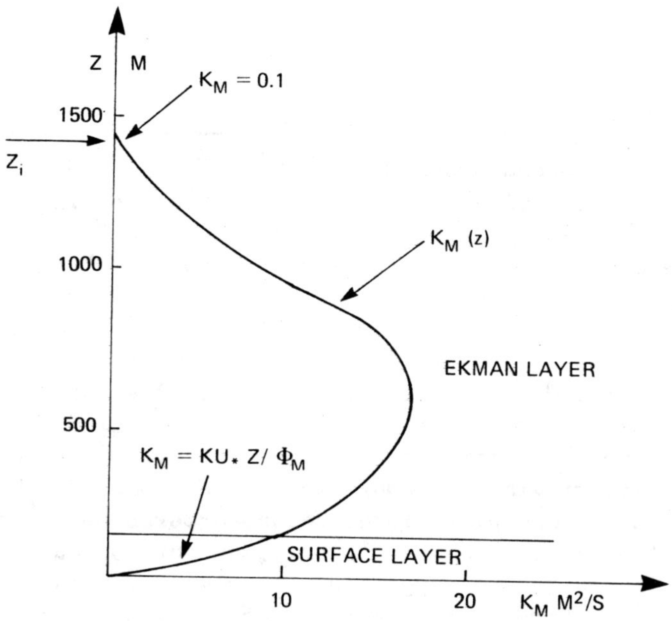

Figure 4.1 Separation of PBL into surface layer and Ekman layer showing typical K_M profile.

4.2.3 Higher Order Closure Schemes

The so called "higher order closure" schemes have, during the last seven-eight years, been successful in simulating many aspects of atmospheric turbulence which were practically unknown a decade ago. Most models of this kind have been devoted to the study of a growing inversion-capped, convective boundary layer driven by an upward heat flux at the surface. Less effort has gone into modelling the stable, nocturnal boundary layer. Wyngaard (1975) belongs to the small group of people having done this.

The higher order schemes are based on deriving prognostic equations for the stresses $\overline{u_i u_j}$ i, J = 1...3. That can be done formally by multiplying the equations of motion, after Egs. (4.2.2) have been introduced, by u_1, u_2 and u_3 successively and then averaging. The result is a set of equations of the form:

$$\frac{\partial(\overline{u_i u_j})}{\partial t} = F \text{ (mean quantities, second-order correlations, triple correlations)}$$

$$\frac{\partial(\overline{w\theta})}{\partial t} = \text{-"-} \qquad (4.2.9)$$

$$\frac{\partial(\overline{wq})}{\partial t} = \text{-"-}$$

If we also include all the second order moments $\overline{\theta^2}$, $\overline{q^2}$, $\overline{u_i\theta}$ and $\overline{u_i q}$ that will give us 14 additional equations to solve. The triple correlation like $\overline{u_i u_j u_k}$ are now terms that have to be specified. By expressing these in some way in terms of the mean quantities and second order correlations a "second order" closure is achieved. By further manipulations it is possible to derive equations for the third order correlations, which contain fourth order terms as unknowns. A closure of the third order has been used by André et al. (1976, I and II). Deardorff (1972, 1973, 1974a, 1974b) has developed a three-dimensional model in which the turbulent fluctuations are resolved on a scale larger than the grid scale and modelled by means of (4.2.9) on the subgrid scale. These simulations have given considerable insight into the behaviour of a convective, growing, boundary layer. Wyngaard and Coté (1974) and Wyngaard (1975) have used a one-dimensional second-order closure model to simulate both the unstable and the stable boundary layer. Wyngaard based his model assumptions on a functional expansion technique suggested by Lumley and Khajeh-Nouri (1974) for expressing the third-order correlations in terms of lower order terms. In Mellor (1973) a second-order model was used for predicting the form of the surface layer functions which was also done in Lewellen and Teske (1973). Mellor and Yamada (1974) developed a series of successive approximations to a second-order closure model with a gradual increase of isotropy in the model approximations. The so called level 3 model is used in a very extensive study of Day 33 of the data (Clarke et al., 1971) by Yamada and Mellor (1975).

Shir (1973) studied an idealized boundary layer and Donaldson (1973) derived a second-order model for air pollution dispersal. Sommeria (1976) has modified Deardorff's (1972) three-dimensional model and used it for simulating the trade wind boundary layer. Schemm and Lipps (1976) have developed a three-dimensional model with simplified subgrid scale correlation equations. In de Moor and André (1975) the higher order modelling is reviewed in somewhat more detail. An instructive discussion is also found in Hossain and Rodi (1974). Deardorff, Wyngaard et al., Yamada and Mellor, André et al., and others have tested their simulation models on the same data set, namely Day 33 of the Wangara experiment in southern Australia. The results for the unstable boundary layer have been very illuminating. Other modellers, using simplified K-theory have also used the same data.

The time requirements on a computer are prohibitive for using higher order closure models in operational applications, especially if three-dimensional models are looked for. On the other hand they generate a multitude of data that can be used for designing more efficient parameterization schemes, for example, K-formulations.

A simplified turbulence closure, but still capable of simulating most of the properties of the PBL with almost the same accuracy as the higher order closure schemes, is the Turbulent Energy equation (TE). In this method only one additional time dependent equation is needed; that for the turbulent energy

$$b = \frac{1}{2}\cdot\{\overline{u^2} + \overline{v^2} + \overline{w^2}\} \qquad (4.2.10)$$

The turbulent energy equation is obtained by contracting the tensor $\overline{u_j u_i}$ to $\overline{u_i^2}$ and adding the corresponding equations (4.2.9).

That will give us for b (assuming horizontal homogeneity)

$$\frac{db}{dt} = - \overline{uw}\frac{\partial U}{\partial z} - \overline{vw}\frac{\partial V}{\partial z} + \frac{g}{\Theta}\overline{w\theta} - \frac{\partial}{\partial z}\left(\overline{wb} + \frac{\overline{pw}}{\rho}\right) - \varepsilon \qquad (4.2.11)$$

By introducing the closure assumptions (4.2.4) and assuming that (Donaldson, 1973) the energy dissipation can be written as

$$\varepsilon = a_3 b^{3/2}/\ell_\varepsilon \qquad (4.2.12)$$

where ℓ_ε is a dissipation mixing length and that (Shir, 1973)

$$-\frac{\partial}{\partial z}\left(\overline{wb} + \frac{\overline{wp}}{\rho}\right) = a_2\cdot\frac{\partial}{\partial z}\left(K\frac{\partial b}{\partial z}\right) \qquad (4.2.13)$$

we get,

$$\frac{\partial b}{\partial t} = K_M\left[S^2 - \frac{g}{\Theta}\cdot\alpha_H\left(\frac{\partial\Theta}{\partial z} - \gamma_{cg}\right)\right] + a_2\frac{\partial}{\partial z}\left(K\frac{\partial b}{\partial z}\right) - \frac{a_3 b^{3/2}}{\ell_\varepsilon} \qquad (4.3.14)$$

where, S is the wind shear defined by (4.2.5)

α_H is the ratio K_H/K_M

γ_{cg} is a correction at unstable stratification which allows for a "counter gradient" heat flux.

Deardorff has suggested the following form for γ_{cg}

$$\gamma_{cg} = 10\ (\overline{w\theta})_s / w_* \cdot z_i \tag{4.2.15}$$

where s refers to surface conditions, z_i is the height of the mixed layer and w_* is a scaling velocity

$$w_* = - \left(g u_* \theta_* z_i / \theta_s\right)^{1/3} \quad , \quad \theta_* \leq 0 \tag{4.2.16}$$

$$= 0 \quad , \quad \theta_* > 0$$

(Deardorff, 1972).

The constant a_2 is of the order of 1.2. The constant a_3 has a value $0.125(2)^{3/2} = 0.35$ according to Donaldson (1973). However, Shir (1973) and Yu (1976) used values of 0.08 and 0.06, respectively.

If (4.2.15) is used in (4.2.14) it should also be included in the turbulent heat flux term in the thermodynamic equation. The final relation

$$K_M = \ell \cdot (0.2b)^{1/2} \tag{4.2.17}$$

closes the system. The main problem is then to specify the mixing lenths ℓ and ℓ_ε and to give α_H.

Equation (4.2.12) has been used by many researchers, especially in the Soviet Union, where it is referred to as Kolmogorov's equation of TE. Bradshaw et al. (1967), were probably the first outside the Soviet Union to use the TE-equation in numerical flow simulations. Petersen (1969, 1972), Gutman (1969), Bodin (1976a,b), Yu (1976), Clarke (1974) and others have used (4.2.14) in numerical models with different assumptions about the mixing length ℓ. Delage (1974) used it in a simulation of the nocturnal boundary layer. Appendix III summarizes different ℓ formulations, which are either extensions of surface layer theory or/and generalizations of Blackadar's (1962) formulation. The shape of ℓ must reflect the dramatic change in scale between stable and unstable stratification and ℓ should reach a value of the order of $1/2 . z_i$ for free convection ($L \to 0$) as is evident from Deardorff (1973). Some formulations fail to do so which inhibits the growth of the convective boundary layer in the model.

The TE-equation makes it possible to incorporate non-stationary effects and PBL-transitions and makes it possible to capture all the essential aspects of the behaviour of turublence in the PBL at a very little extra cost as compared with K-theory. The TE-approach also has a stabilizing effect on the numerical schemes commonly in use in PBL-models.

4.3 Operational Models

In applying boundary layer models we can either specialize in solving a particular problem, for example, the prediction of fog, or give actual forecast information about the structure of the whole PBL. The latter aspect is becoming of great interest to weather services that want to rationally use the output from synoptic scale numerical weather prediction models. Boundary layer models can then be used as interpretative tools for making local forecasts. This problem was discussed at the WMO Symposium on the Interpretation of broad-scale NWP products (Warsaw, October, 1976). Models that are going to be used in such operational environments must be capable of efficiently handling all important types of boundary layer processes. Naturally computer capacity puts limits to what can be achieved in that respect, but a one-dimensional PBL-model might give useful information for places where assumptions of horizontal homogeneity can be approximately fulfilled. Of course, three-dimensional PBL-models have to be used in order to introduce the effects of varying surface conditions. What then are the requirements on such models?

4.3.1 Physical and Computational Requirements

The physical processes we need to model are both being controlled or forced (over a timescale of about 24h) by the large scale, non-stationary, synoptic flow and influenced by local parameters such as roughness, topography and the heat and moisture properties of the surface.

Figure 4.2 illustrates the different parameters and processes we have to know when modelling the PBL. Many of these processes are neglected in the otherwise sophisticated simulation models but have to be included in some way in operational models.

(a) Turbulence - is of prime concern in a PBL-model. See above.

(b) Radiation - both long-wave and short-wave has to be computed by the model. This is a very time-consuming part of a PBL-model. Different methods of calculating the radiative heat flux divergences are in use. Papers dealing with these problems are: Atwater (1966, 1970), Atwater and Brown (1974), Kondratyev (1969), Kuhn (1963), McDonald (1960) and Zdunkowski and Johnson (1965). Zdunkowski and Barr (1972) and Brown and Roach (1976) treat the problem of radiation in the presence of fog, or cloud droplets, in the PBL. Shaffer and Long (1973) describe a complete radiation package, based on Atwater (1970), suitable for PBL-models. Welch and Zdunkowski (1976) present a radiation model for polluted PBL's.

(c) Condensation - is very important in PBL-models and is often one of the goals of the model, e.g. to predict fog and cloud formation.

The phase changes of water vapour are incoporated in different ways, most of which are based on linearizations of moist adiabatic processes for shallow convection. Betts (1974) discusses the use of conservation of static energy suggested by Madden and Robitaille (1970) while Deardorff (1976) has demonstrated that using a liquid-water potential temperature as suggested by Betts (1973) gives numerical advantages. Brown and Roach (1976) deal with the rate of settlements of fog droplets and its effect on liquid water content in fogs. Condensation changes the radiative and turbulent properties of the PBL.

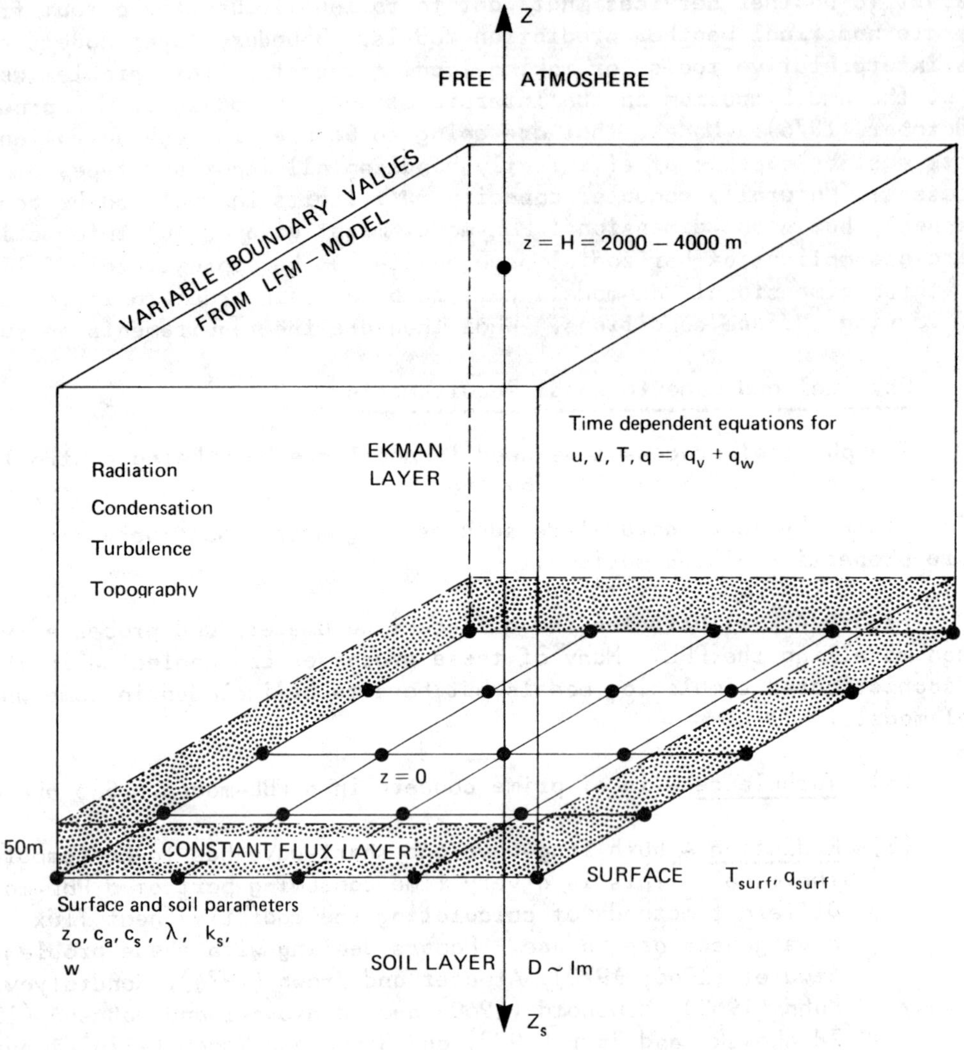

Figure 4.2 Schematic structure of an Atmospheric Boundary Layer Model. z_0 roughness parameter, c_a, c_s specific heat for air and soil respectively, λ thermal conductivity and k_s thermal diffusivity in the soil. w is soil water content.

(d) Soil Heat and Soil Moisture - In order to include a surface temperature prediction to provide the PBL model with lower boundary conditions, it is necessary to use some kind of energy balance at the surface, which requires a knowledge of soil heat flux and evaporation of water from the surface. We know very little about moisture in the ground and evapotranspiration from the vegetation. These problems are discussed in many papers. Most important are Deardorff (1977), which has an extensive list of references to this subject, De Vries (1973) and Jacobs and Brown (1973). Zdunkowski et al. (1975) and Sasamori (1970) use a quite complete soil moisture model. Nappo (1975) compares different lower boundary formulations and Zdunkowski and Trask (1971) make comparison of the influence of different soil types.

(e) Topography - has a strong effect on the boundary layer. This is also true for spatial variations in z_o as many theoretical and numerical investigations have shown. In most real life situations the flow encounters numerous changes in surface properties, which would require extremely fine horizontal resolution to resolve.

(f) Non-stationarity and baroclinicity - Many theoretical models employ assumptions of steady state and horizontal homogeneity. In reality there are large deviations from this ideal state and different similarity assumptions are of limited value. The PBL is often baroclinic in contrast to the assumed barotropy. Models based on the Ekman-type of equations (non-stationary) gives rise to false inertial-diffusive oscillations as demonstrated by Ching and Businger (1968). They have a period of a pendulum day and can hardly be filtered by traditional technique. As demonstrated in Bodon (1974), these false oscillations are, however, eliminated by using Gutman's approach to mesoscale and boundary layer models (Gutman, 1969; Gutman and Perov, 1970).

4.3.2 Scale Analysis of the Equations of Motions and the Question of One-Dimensional or Three-Dimensional Models

Barr and Kreitzberg (1975) argue that in a three-dimensional boundary layer model with a horizontal grid distance of $\Delta s \sim 100$ km or more each grid-point will appear as an isolated one-dimensional model because of weak horizontal coupling as compared to the vertical exchange. In Bodin (1976b) a scale analysis was carried out of the pertinent equations of motion. It was demonstrated that the turbulent exchange term characterized by the Ekman number, is of the order of 10^0. On the other hand the horizontal advections in the boundary layer depend on the horizontal scale through a "boundary layer" Rossby number

$$Ro_b = \frac{u_o}{fL_b} \sim 10^5 \cdot 10^{-x} \qquad (4.3.1)$$

if we put $L_b \simeq 10^x$.

If we assume

$$L_b = 10^3 \text{ km}(10^6 \text{m}) \Rightarrow Ro_b \sim 10^{-1}$$

or

$$L_b = 100 \text{ km} \Rightarrow Ro_b \sim 10^0$$

we see that when we go down to scales of the order of 100 km the horizontal advections have the same magnitude as the vertical diffusion. At $L_b \sim 1000$ km (correspondent to a grid distance of ~ 30 km) Ro_b is the one order of magnitude less than the corresponding Ekman number. This leads to the conclusion that it is only profitable to have a three-dimensional boundary layer model if we can afford a grid distance of the order of 30 km, say 10 km. Otherwise it is better to stay one-dimensional.

4.3.3 Numerical Methods and Finite Differences

Most problems in numerical boundary layer modelling reduce to solving one or more equations of the type

$$\frac{\partial Q}{\partial t} = \frac{\partial}{\partial z}\left(K \frac{\partial Q}{\partial z}\right) + F \qquad (4.3.2)$$

where F represents low order terms. Equation (4.3.2) is the typical parabolic diffusion equation with the property that K varies considerably through the domain. Furthermore K is some function of Q itself making (4.3.2) non-linear. Standard schemes for solving (4.3.2) are found for example Richtymer and Morton (1967). Long (1975) discussed at some length various numerical schemes and their relative merits in terms of a computational diffusion coefficient K_c. The use of different coordinate transformations to increase the accuracy close to the ground is discussed in Taylor and Delage (1971) and Delage (1974) and Brook (1975). Other papers relevant to the diffusion equation are Estoque and Bhumralkar (1970), Lilly (1965) and Paegle et al. (1976). Most experiences indicate that the Crank-Nicholson (or the Laasonen) scheme is the most appropriate one for boundary layer problems. Horizontal advection schemes are discussed in Long and Pepper (1976), Lilly (1965), Le Blanc (1968), Bornstein and Robock (1976) and Long (1975). A pseudo-spectral technique for a limited domain has also been developed by Christensen and Prahm (1976).

4.3.4 Brief Description of Some Models

In this section we will discuss some specific, recent, models. The models of the 1960's represented substantial progresses in boundary layer modelling but suffered from the lack of developed turbulence parameterization. In this group belong the models of Estoque (1963), Pandolfo et al. (1965) and Sasamori (1970). A variety of K-models have been used in numerous air pollution models. In the 1970's more elaborate and physically realistic K-models have been put into use. The experiences from higher order closure simulation models have led to improvements like the use of the TE-equation and simplified second order models like Yamada and Mellor's, which are well within reach for operational one-dimensional models.

Models of the "Gutman" kind:

Gutman (1969) suggests a method of formulating mesoscale and boundary layer models based on a subtraction of the large scale flow from the boundary layer equations. This approach has been applied in a series of papers: Gutman and Perov (1970), Lykosov (1972), Lykosov and Gutman (1972) and Speranskiy et al. (1975).

The Gutman approach is based on splitting up the variables into two parts, for example the wind field

$$u = u_L + u_b$$

$$v = v_L + v_b \qquad (4.3.3)$$

where index L refers to large scale flow and b to boundary layer flow. By inserting (4.3.3) in the Reynold's equations and then subtracting the equations for the large scale flow and neglecting horizontal advections we arrive at equations for $\frac{\partial u_b}{\partial t}$ and $\frac{\partial v_b}{\partial t}$. This procedure is especially beneficial if the goal is to couple the PBL-model to a large scale NWP-model.

In that case the equations for the large scale "background" flow are simply given by the primitive equations used.

The Gutman approach has three advantages when treating non-stationary problems (i.e. variable boundary values in time).

(a) The absence of the gestrophic wind in the equations filters out the class or (false) inertial-diffusive oscillations discussed by Ching and Businger (1968) and Bodin (1974).

(b) No need to specify both the geostrophic wind and the real wind at the upper boundary.

(c) Implicit incorporation of the effect of large scale advections.

The Gutman approach has been used in Bodin (1976,a,b). Bodin's model uses the TE-equation as in (4.2.14) with (4.2.15) and (4.2.17). The mixing length formulation is that of Eqs. (6) and (7) in Appendix II. The numerical solution makes use of a coordinate transformation in the vertical and the Laasonen scheme for the time integration. Time step is four minutes and 35 gridpoints between the surface and the top at 2000 m are used. A rate equation for the surface temperature is used taking into account radiation and surface soil heat flux. The model has been tested on Day 33 of the Wangara data (Clarke et al., 1971) and Finnish data with very good results for both the night-time and daytime boundary layer. Gutman's model has been tested on the O'Neill data (Lettau and Davidson, 1957) by Speranskiy et al. (1975).

Yamada and Mellor's model:

Yamada and Mellor (1975) presented a simulation of Day 33, 34 and 35 of the Wangara data. The model is an application of the "level 3" model derived in Mellor and Yamada (1974).

In addition to the TE-equation Yamada and Mellor also carry equations for the temperature variance, $\overline{\theta^2}$ and moisture variance $\overline{q^2}$. Time integration is carried out by means of Laasonen's scheme with $\Delta t = 60$ s. The vertical coordinate is transformed in a similar way as in Bodin. Eighty points are used between the ground and 2000 m. In the model the effect of large scale subsidence and vertical variation of the geostrophic wind, i.e. baroclinicity, are accounted for. Boundary conditions are prescribed from observed screen measurements of temperature and specific humidity. Both the wind and the temperature profiles show good agreement with observations. Some deficiencies might be removed by the use of a different formulation of the mixing length. Burk (1977) has used Yamada and Mellor's level 3 model to study the "moist" boundary layer.

Other models:

For the sake of completeness some other models should also be mentioned. These include the one-dimensional model of Busch and Chang (1976), Yu (1976), Karlsson (1972), and Karlsson and Aspling (1977), Geleyn (1975) and Walmsley (1973), in which air-sea interaction is studied. Some special application models are devoted to fog-prediction, for example, Lala et al. (1975), Brown and Roach (1976), Kessler and Baker (1976), Zdunkowski and Barr (1972) and Barker (1977). Smedman-Högström (1976) deals with the problem of predicting runway icing.

Three-dimensional models:

The first attempt to develop an operational three-dimensional model was carried out by Gerrity (1967). That model was intended for the prediction of low ceiling and cloudiness. The model was later taken over by U.S. Air Force and implemented in an operational contest at USAF Global Weather Centre as reported by Hadeen and Friend (1972) and Hadeen (1970). The model has also been tested with some modifications at the National Meteorological Centre (NMC) in Washington, D.C., as reported by Gerrity et al. (1972) and Gross et al. (1972).

In Polger (1974) an attempt is made to evaluate the predictive power of the model. Even though the model showed a certain skill in predicting, for example, the type of precipitation its performance was not judged as good enough to run operationally. Instead NMC has turned to another three-dimensional model, not basically different from Gerrity's but more up-to-date. This second model is being developed at Techniques Development Laboratories (TDL) by Shaffer, Long, Kemper and Yu.

The three-dimensional models of Padro (1974), Shaffer and Long (1973) and Long and Shaffer (1975) use similar assumptions as Gerrity's (1967). The requirements for fast parameterization become very important when using three-dimensional models for operational applications and sacrifices have to be made in physical fidelity.

A natural way, followed by Gerrity, is to eliminate the computations in the lowest 50 meters by means of a constant flux layer or surface layer as suggested in Eqs. (4.2.6–4.2.7). Gerrity used a steady state solution for the Ekman layer using the values of K at the top of the surface layer from Eqs. (4.2.8) and keeping them constant up to 1600 m. Temperature, T, and specific humidity, Q were predicted from time dependent equations, including non-adiabatic effects and turbulent diffusion.

The forecast equations were discretized over an area resolved by 29 x 27 grid-points with a separation of 190.5 km. Vertically the PBL was divided into seven layers from the ground to 1600 m where the upper boundary conditions for V_g, θ and Q are supplied from the Limited-area Fine-mesh Model (LFM). Gerrity used the Crank-Nicholson method for the diffusion and vertical advection terms and upstream differencing for the horizontal advection terms. This allowed a time step of 15 minutes. A weak point in the model was a very simple treatment of surface temperature and moisture at the lower boundary. Radiation was also simulated in a simple, climatological, way to reduce the integration time. Shaffer and Long's model includes a considerable extension of Gerrity's model with a more refined and complete treatment of the physics in the PBL.

Padro's model (1974), developed in Canada, is half way between Gerrity and Shaffer and Long. He uses Businger's surface relations and extends the corresponding K through the Ekman layer by means of the third order O'Brien interpolation polynomial (see Appendix I). The winds in the Ekman layer are computed diagnostically from steady state Ekman layer equations as in Gerrity, but with the variable O'Brien K-profile. The grid distance is 127 km and Δt = 30 minutes. Both Padro and Gerrity use a terrain following height coordinate system.

Shaffer and Long's Model:

In Shaffer and Long (1973), Long and Shaffer (1975) and Long (1975) various aspects of the development of a three-dimensional boundary layer model are discussed.

Shaffer and Long (hereafter referred to as SL) write the equations of motion in complex form to make efficient use of the complex arithmetic on the IBM 360/195 and

$$\frac{\partial \tilde{w}}{\partial t} + \underline{V} \cdot \nabla \tilde{w} = \frac{\partial}{\partial z} \left(K_M \frac{\partial \tilde{w}}{\partial z}\right) - i \left[\underline{f} \, (\tilde{w} - \tilde{G})\right] \qquad (4.3.4)$$

$$\frac{\partial \tilde{T}}{\partial t} + \underline{V} \cdot \nabla \tilde{T} = \frac{\partial}{\partial z} \left(K_H \frac{\partial \tilde{T}}{\partial z}\right) + \tilde{Q} \qquad (4.3.5)$$

where $\tilde{w} = U + i \cdot V$, $\tilde{G} = U_g + i \cdot V_g$, $\tilde{T} = \Theta + i \cdot Q$

$\tilde{Q} = Q_h + i \cdot Q_q$ and $K_H = K_E \neq K_M$

Q_h and Q_q are source terms for heat and moisture respectively. In the surface layer Businger's flux profiles are integrated and in the Ekman layer an O'Brien profile of K is used.

To solve for the surface temperature a surface energy balance equation is used.

$$\rho c_p u_* \theta_* + \rho L u_* q_* - F_{soil} - \varepsilon \sigma T^4 + \text{Radnet} = 0 \qquad (4.3.6)$$

where the two first terms are the fluxes of sensible and latent heat.

The evoporation is calculated from assuming a constant ratio between actual to potential evaporation (the so called Halstead parameter) $H = E/E_p = (\varrho_h - \varrho_{z=0})/(\varrho_h - \varrho_{z=0}(\text{saturated}))$, where $H = f(w)$ is a function of soil water water content, Nappo (1975).

In (4.3.6) the soil heat flux also has to be given. The usual procedure is to solve the heat conduction equation for a soil layer to give surface soil heat flux. By assuming a constant thermal diffusivity, K_s, and conductivity, λ_s, the simple heat conduction equation can be solved analytically for the boundary conditions

$$
\begin{aligned}
T_{soil} &= T_{oo} \qquad z \to oo \\
T(z) &= T_o(z) \qquad t = 0
\end{aligned}
\tag{4.3.7}
$$

and

$$
T_{sfc} = \frac{1}{\sqrt{\pi k_s t}} \int_0^{\infty} T_o(\zeta) e^{-\zeta^2/4k_s t} \, d\zeta + \frac{1}{\lambda_s} \sqrt{\frac{k_s}{\pi}} \int_0^t \frac{F(t-u)}{\sqrt{u}} \, du \tag{4.3.8}
$$

By approximating the second integral with finite differences the surface soil heat flux F can be solved for in terms of previous surface temperatures. This gives a rapid calculation of the soil heat fluxes. A similar method was used by Gutman (1953). Bodin (1974) has shown that (4.3.8) gives almost the same accuracy as a detailed numerical solution of the heat conduction equation.

SL's model has a complete calculation of short and long wave radiation developed from Atwater. It uses emissivity functions by Kuhn (1963) and CO_2 emissivities by Kondratyev (1969). Clouds are accounted for in an approximate way. Finite differences are discussed in Long (1975). The physics of a one-dimensional version of the model have been tested on Wangara and O'Neill data.

In the three-dimensional version the surface layer is 50 m thick and eight equidistant levels are used up to 2000 m. The model is developed for a domain containing 35 x 30 grid-points 80 km apart. This allows a time step of 1/2 hour. The model is presently being tested on real data.

Sea breeze models:

Two three-dimensional models developed primarily for the simulation of sea breezes are of particular interest since they both can be generalized and used operationally.

Pielke's model:

This model has been used to simulate the sea breeze over Florida as reported in Pielke (1974a,b). The latest development is described in Pielke and

Mahrer (1975). The model structure is not very different from SL. In the (1975) paper an improvement is made by introducing a rate equation for the boundary layer height z_i suggested by Deardorff.

Neglecting the influence of clouds

$$\frac{\partial z_i}{\partial t} = - V \cdot \nabla z_i + w_h + \frac{1.8\left(w_*^3 + 1.1u_*^3 - 3.3u_*^2 f z_i\right)}{\frac{g z_i^2}{\theta_s}\frac{\partial \theta}{\partial z}^+ + 9.0w_*^2 + 7.2u_*^2} \qquad (4.3.9)$$

where w_* is defined in (4.2.16) and $\frac{\partial \theta}{\partial z}^+$ is the lapse rate above the capping inversion and w_h is the large scale vertical velocity at the top of the boundary layer. By putting the top of the O'Brien profile at z_i instead of at some fixed height the prediction of a growing convective boundary layer is considerably improved at a very low "cost". A similar modification should also be possible for the nocturnal boundary layer. The problem, however, is more complicated due to the presence of patchy turbulence induced by breaking gravity waves.

Tapp and White's model:

This model, developed at the British Meteorological Office, is described in a recent paper by Tapp and White (1976). The main difference from other models is that the model is non-hydrostatic and allows the propogation of sound waves. Tapp and White discuss at some length the implications of the presence of sound waves and the final finite difference scheme used. In order to allow reasonably long time steps a semi-implicit integration method is used. But even by doing so a time step of 60 seconds has to be used. But as a three-dimensional simulation model it has certain advantages. The model has been applied both to sea breezes over the British Isles as well as compared with Pielke's model. The results have been favourable and there seem to be small differences between the non-hydrostatic and hydrostatic model when similar resolution is used.

4.4 Data Requirements and Initialization

Boundary layer models depend on a number of ill-defined physical parameters like thermal conductivity in the soil, soil water content and roughness height. Compromises have to be made between numerical efficiency and modelling of the physical processes in the atmosphere, in the ground and at the interface which is frequently a layer of vegetation. Radiation is the most time consuming part of a model but it supplies the input of energy. There exist fairly fast and accurate parameterization methods for the non-disturbed boundary layer but very little is known of the effects of large scale advections, non-stationarity, interactions with Cu- and Cb-clouds, rain falling through the boundary layer, etc. In addition the data available for an operational model are not from observations aimed at studying the boundary layer.

The question is what kind of data do we need in order to produce meaningful information about the PBL? In most countries we have to rely on the convectional net of surface observations and radiosonde data, which are designed for quite another purpose. J. Kemper at TDL has been experimenting with a variational scheme to analyze convectional data and initialize the TDL model. Initial inbalances might set up false diffusive-inertial oscillations which are almost impossible to filter out by ordinary techniques. Sasaki and Lewis have earlier (1970) used the variational method in PBL-analysis.

Kreitzberg (1976) has recently reviewed the possibilities in getting data by future satellite systems for mesoscale models. There might even be possibilities for estimating soil moisture content as reported by Isdo et al. (1976).

Simple experiments with the already existing models are required to analyse the needs of initial data for PBL-models. It is in this context, of considerable importance to be able to design correctly the improvements in the present data collection systems which will be necessary to produce information about the PBL beneficial to a large number of users and to the weather services.

References

André, J.C., G. de Moor, P. Lacarrère and R. Du Vachat, 1976: Turbulence approximation for inhomogeneous flows: Part I. The clipping approximation. Part II. The numerical simulation of a penetrative convection experiment. *J. Atmos. Sci.*, 33, 476-491.

Atwater, M.A., 1966: Comparison of numerical methods for computing radiative temperature changes in the atmospheric boundary layer. *J. Appl. Meteor.*, 5, 824-831.

Atwater, M.A., 1970: Investigation of the radiation balance for polluted layers of an urban environment. Ph.D. Dissertation, New York Univ., New York, 116 pp.

Atwater, M.A. and P.S. Brown Jr., 1974: Numerical computation of the latitudinal variation of solar radiation for an atmosphere of varying capacity. *J. Appl. Meteor.*, 13, 289-297.

Barker, E.H., 1977: A maritime boundary layer model for the prediction of fog. *Boundary-Layer Meteor.*, 11, 267-294.

Barker, E.H. and T.L. Baxter, 1975: A note on the computation of atmosphere surface fluxes for use in numerical modeling. *J. Appl. Meteor.*, 14, 620-622.

Barr, A.E. and C.W. Kreitzberg, 1975: Horizontal variability and boundary layer modelling. *Boundary-Layer Meteor.*, 8, 163-172.

Betts, A.K., 1973: Non-precipitating cumulus-convection and its parameterization. *Quart. J.R. Meteor. Soc.*, 99, 178-196.

Betts, A.K., 1974: Further comments on "A comparison of the equivalent potential temperature and the static energy". *J. Atmos. Sci.*, 31, 1713-1715.

Blackadar, A.K., 1962: The vertical distribution of wind and turbulent exchange in a neutral atmosphere. *J. Geophys. Res.*, 67, 3095-3102.

Bodin, S., 1974: Development of an unsteady atmospheric boundary layer model. SMHI-Reports, Meteorology and Climatology, No. 2.

Bodin, S., 1976a: Numerical models of the boundary layer. Seminar on the Treatment of the Boundary Layer in NWP. ECMWF. Reading, 6-10, Sept. 1976.

Bodin, S., 1976b: An unsteady one-dimensional atmospheric boundary layer model. Proceedings from the WMO Symposium on the Interpretation of Broadscale NWP Products for Local Forecasting Purposes. Warsaw, October 1976. WMO-No. 450.

Bodin, S., 1977: Simple steady and unsteady models of the atmospheric boundary layer - A review prepared for Nordforsk. Annual Report 1976 - Nordforsk - Mesoscale dispersion models.

Bornstein, R.D. and A.D. Robock, 1976: Effects of variable and unequal time step for advective and diffusive processes in simulations of the urban boundary layer. *Mon. Wea. Rev.*, 104, 260-267.

Bradshaw, P., D.H. Ferris and N.P. Atwell, 1967: Calculation of boundary layer development using the turbulent energy equation. *J. Fluid Mech.*, 28, 593-616.

Brook, R.R., 1975: Rationale for determining spacing of levels in numerical models of the atmospheric boundary layer. *Boundary-Layer Meteor.*, *Vol.* 8, No. 3/4.

Brown, R. and W.T. Roach, 1976: The physics of radiation fog: II - A numerical study. *Quart. J.R. Meteor. Soc.*, 102, 335-354.

Busch, N. and S.W. Chang, 1976: A multilevel model of the planetary boundary layer suitable for use with meso-scale dynamic models. Proceedings from ECMWF Seminars on the Treatment of the Boundary Layer in Numerical Weather Prediction. Reading, September 1976.

Businger, J.A., J.C. Wyngaard, Y. Izumi and E.F. Bradley, 1971: Flux-profile relationships in the atmospheric surface layer. *J. Atmos. Sci.*, 28, 181-189.

Ching, J.K.S. and J.A. Businger, 1968: The response of the planetary boundary to time varying pressure gradient force. *J. Atmos. Sci.*, 25, 1021-1025.

Clarke, R.H., A.J. Dyer, R.R. Brook, D.G. Reid and A.J. Troup, 1971: The Wangara Experiment: Boundary Layer Data. CSIRO, Division of Meteorological Physics Techn. Paper No. 19.

Clarke, R.H., 1974: Attempts to simulate the diurnal course of meteorological variables in the boundary layer. *Izvestiya, Atm. and Oceanic Physics*, 10, (6), (English translation), 360-374.

Christensen, O. and L.P. Prahm, 1976: A pseudo-spectral model for dispersion of atmospheric pollutants. *J. Appl. Meteor.*, 15, 1284-1294.

Deardorff, J.W., 1972: Numerical investigations of neutral and unstable planetary boundary layers. *J. Atmos. Sci.*, *29*, 91-115.

Deardorff, J.W., 1973: The use of subgrid transport equations in a three-dimensional model of atmospheric turbulence. Paper presented at the Applied Mechanics and Fluid Engineering Conference, Atlanta, Ga., June 1972.

Deardorff, J.W., 1974a: Three-dimensional numerical study of turbulence in an entraining mixed layer. *Boundary-Layer Meteor.*, *7*, 199-226.

Deardorff, J.W., 1974b: Three-dimensional numerical study of the height and mean structure of a heated planetary boundary layer. *Boundary-Layer Meteor.*, *7*, 81-106.

Deardorff, J.W., 1976: Usefulness of liquid-water potential temperature in a shallow-cloud model. *J. Appl. Meteor.*, *15*, 98-102.

Deardorff, J.W., 1977: Efficient prediction of ground surface temperature and moisture with inclusion of a layer of vegetation. NCAR-Research Papers, January 1977.

Delage, Y., 1974: A numerical study of the nocturnal atmospheric boundary layer. *Quart. J.R. Meteor. Soc.*, *100*, 351-364.

Djolov, G.D., 1973: Modeling of interdependent diurnal variation of meteorological elements in the boundary layer. Ph.D. Thesis, University of Waterloo, Canada.

Donaldson, C. du P., 1973: Construction of a dynamic model of atmospheric turbulence and the dispersal of atmospheric pollutants. (D.A. Haugen, ed.). Workshop in Micrometeorology, American Met. Soc., Boston, 392 pp.

de Moor, G. and J.C. André, 1975: La turbulence dans la couche limite atmosphérique. La Meteorologie, Series 7, No. 3, 179-198.

De Vries, D.A., 1975: Heat transfer in soils. Heat and mass transfer in the biosphere. Scripta Book Company. John Wiley & Sons, New York.

Estoque, M.A., 1963: A numerical model of the atmospheric boundary layer. *J. Geophys. Res.*, *68*, 1103-1113.

Estoque, M.A. and C.M. Bhumralkar, 1970: A method for solving the planetary boundary layer equations. *Boundary-Layer Meteor.*, *1*, 169-194.

Geleyn, J.F., 1975: Un modele unidimensionel et son application a l'étude du cycle diurne. La Meteorologie, Series 7, No. 2, 17-34.

Gerrity, J.P., 1967: A physical-numerical model for the prediction of synoptic-scale cloudiness. *Mon. Wea. Rev.*, 95, 261-282.

Gerrity, J.P. et al., 1972: On the feasibility of integrating a combined LFM-PBL model. Office Note 77, National Meteorological Center, NOAA, USA.

Gross, E., R. Jones and R. McPherson, 1972: A description of the NMC planetary boundary layer model. Office Note 75, National Meteorological Center, NOAA, USA, 50 pp.

Gutman, L.N., 1953: A note on the calculation of heat state of solid bodies. Akad. Nauk USSR, Inst. Mechanics, TR., 15, 99 p.

Gutman, L.N., 1969: Introduction to the non-linear theory of meso-scale processes. Leningrad 1969. English translation. Israel Programme for Scientific Translations, Jerusalem, 1972.

Gutman, L.N. and V.L. Perov, 1970: Non-stationary problems in meso meteorology. *Izv. Atm. and Oceanic Phys.*, 6, (English translation), 1-6.

Hadeen, K.D., 1970: AFGWC Boundary Layer Model. AFGWC Tech. Memo 70-5, Air Force Global Weather Central, Offut AFB, Neb., USA.

Hadeen, K.D. and A.L. Friend, 1972: The Air Force Global Weather Central operational boundary-layer model. *Boundary-Layer Meteor.*, 3, 98-112.

Hossain, M.S. and W. Rodi, 1974: Equations for turbulent buoyant flows and their modelling. Sonderforschungsbereich 80, Ausbreitungs-und Transportvorgänge in Strömmungen. Rep. SFB 80/T/46. Universität Karlsruhe.

Idso, S.B., R.D. Jackson and R.J. Reginato, 1976: Compensating for environmental variability in the thermal inertia approach to remote sensing of soil moisture. *J. Appl. Meteor.*, 15, 811-817.

Jacobs, C.A. and P.S. Brown Jr., 1973: An investigation of the numerical properties of the surface heat-balance equation. *J. Appl. Meteor.*, 12, 1069-1072.

Karlsson, E., 1972: A numerical model for the boundary layer of the atmosphere at neutral and stable stratification. DM-7, Inst. of Meteorology, University of Stockholm. (In Swedish).

Karlsson, E. and G. Aspling, 1977: Experiments with a model for the boundary layer of the atmosphere. MVC Report F8, Swedish Military Weather Service.

Kessler, R.C. and D.D. Baker, 1976: Application of a numerical fog prediction model to radiosonde data. Paper presented at AMS 6th Conference on Weather Forecasting and Analysis, May 1976, Albany, New York.

Kondratyev, K. Ya., 1969: Radiation in the Atmosphere. International Geophysics Series, 12, Academic Press, New York.

Kreitzberg, C.W., 1976: Interactive applications of satellite observations and meso-scale numerical models. *Bull. of AMS*, 57, 679-685.

Kuhn, P.M., 1963: Radiometersonde observations of infrared flux emissivity of water vapor. *J. Appl. Meteor.*, 2, 368-378.

Lala, G.G., E. Mandel and J.E. Jiusto, 1975: A numerical evaluation of radiation fog variables. *J. Atmos. Sci.*, 33, 720-728.

LeBlanc, L.L., 1968: Finite-difference methods used in models of the atmospheric boundary layer. Air Weather Service (MAC), USAF, Technical Rep. 206.

Lettau, H.H. and B. Davidson, 1957: Exploring the atmosphere's first mile, Vol. 2, New York, Pergamon Press.

Lewellen, W.S. and M. Teske, 1973: Prediction of the Monin-Obukhov similarity functions from an invariant model of turbulence. *J. Atmos. Sci.*, 30, 1340-1345.

Lilly, D.K., 1965: On the computational stability of numerical solutions of time-dependent non-linear geophysical fluid dynamics problems. *Mon. Wea. Rev.*, 93, 11-26.

Long, P.E., 1975: Dissipation, dispersion and difference schemes. NOAA Tech. Memo. NWS TDL-55.

Long, P.E. and P.W. Pepper, 1976: Comparison of six numerical schemes for calculating the advection of atmospheric pollution. AMS 3rd Symposium on Atm. Turbulence, Diffusion and Air Quality, Raleigh, N.C., October 1976.

Long, P.E. and W.A. Shaffer, 1975: Some physical and numerical aspects of boundary layer modeling. NOAA Tech. Memo. NWS TDL-56.

Lumley, J.L. and B. Khajeh-Nouri, 1974: Computational modeling of turbulent transport. *Advances in Geophysics*, 18A, 169-192.

Lykosov, V.N., 1972: Unsteady state in the planetary boundary layer of the atmosphere. *Izv. Atmos. and Ocean. Phys.*, 8, (2), (English translation), 85-91.

Lykosov, V.N. and L.N. Gutman, 1972: Turbulent boundary layer above a sloping underlying surface. *Izv. Atm. and Ocean. Phys.*, 8, (8), (English translation), 462-467.

Madden, R.A. and F.E. Robitaille, 1970: A comparison of the equivalent potential temperature and the static energy. *J. Atmos. Sci.*, 27, 327-329.

McDonald, J.E., 1960: Direct absorbation of solar radiation by atmospheric water vapor. *J. Meteor.*, 17, 319-328.

Mellor, G.L., 1973: Analytic prediction of the properties of stratified planetary surface layers. *J. Atmos. Sci.*, 30, 1061-1069.

Mellor, G.L. and T. Yamada, 1974: A hierarchy of turbulence closure models for planetary boundary layers. *J. Atmos. Sci.*, 31, 1791-1806.

Nappo, C.J. Jr., 1975: Parameterization of surface moisture and evaporation rate in a planetary boundary model. *J. Appl. Meteor.*, 14, 289-296.

Nickerson, E.C. and V.E. Smiley, 1975: Surface layer and energy budget parameterizations for mesoscale models. *J. Appl. Meteor.*, 14, 297-300.

O'Brien, J.J., 1970: A note on the vertical structure of eddy exchange coefficient in the planetary boundary layer. *J. Atmos. Sci.*, 27, 1213-1215.

Padro, J., 1974: Forecasting in the planetary boundary layer. Preliminary Report, Atmospheric Environment Service, Canada.

Pandolfo, J.P. et al., 1965: The development of a numerical prediction model for the planetary boundary layer. Report 7465-174. Travelers Research Center. U.S. Weather Bureau Contract CWB-10960. Final Report.

Paulson, C.A., 1970: The mathematical representation of wind speed and temperature profiles in the unstable atmospheric surface layer. *J. Appl. Meteor.*, 9, 857-861.

Peagle, J., W. Zdunkowski and R. Welch, 1976: Implicit differencing of predictive equations of the boundary layer. *Mon. Wea. Rev.*, 104, 1321-1324.

Petersen, E.W., 1969: Modification of mean flow and turbulent energy by a change in surface roughness under conditions of neutral stability. *Quart. J. R. Meteor. Soc.*, 95, 561-575.

Petersen, E.W., 1972: Relative importance of terms in the turbulent energy and momentum equations as applied to the problem of a surface roughness change. *J. Atmos. Sci.*, 29, 1470-1476.

Pielke, R.A., 1973: A three-dimensional numerical model of the sea breezes over south Florida. NOAA Tech. Memo. ERL WMPO-2, 136 pp.

Pielke, R.A., 1974a: A three-dimensional numerical model of the sea breezes. *Mon. Wea. Rev.*, 102, 115-139.

Pielke, R.A., 1974b: A comparison of three-dimensional and two-dimensional numerical predictions of sea breezes. *J. Atmos. Sci.*, 31, 1577-1585.

Pielke, R.A. and Y. Mahrer, 1975: Representation of the heated planetary boundary layer in meso-scale models with coarse vertical resolution. *J. Atmos. Sci.*, 32, 2288-2308.

Polger, P.D., 1974: An evaluation of the National Meteorological Center's experimental boundary layer model. NOAA Tech. Memo. NWS NMC-55.

Prandtl, L., 1925: Bericht über Untersuchungen zur ausgebildeten Turbulenz. *Angew. Math. Mech.*, 5, 136.

Richtmyer, R.D. and K.W. Morton, 1967: Difference methods for initial-value problems. Interscience Publisher.

Sasaki, Y. and J.M. Lewis, 1970: Numerical variational objective analysis of the planetary boundary layer in conjunction with squall line formation. *J. Meteor. Soc., Japan*, 48, 381-399.

Sasamori, T., 1970: A numerical study of atmospheric and soil boundary layers. *J. Atmos. Sci.*, 27, 1122-1137.

Shaffer, W.A. and P.E. Long, 1973: A predictive boundary layer model. Paper presented at the Symposium on the Atmospheric Boundary Layer, October 1973, Mainz, Germany.

Schemm, C.E. and F.B. Lipps, 1976: Some results from a simplified three-dimensional numerical model of atmospheric turbulence. *J. Atmos. Sci.*, 33, 1021-1041.

Shir, C.C., 1973: A preliminary numerical study of atmospheric turbulent flows in the idealized planetary boundary layer. *J. Atmos. Sci.*, 30, 1327-1339.

Shir, C.C. and R.D. Bornstein, 1977: Eddy exchange coefficients in numerical models of the planetary boundary layer. *Boundary-Layer Meteor.*, 11, 171-186.

Smedman-Högström, A.S., 1976: A method to predict icing on runways. Report, Inst. of Meteor., University of Uppsala.

Sommeria, G., 1976: Three-dimensional simulation of turbulent processes in an undisturbed trade wind boundary layer. *J. Atmos. Sci.*, 33, 216-241.

Sorbjan, Z., 1976: Parameterization of dynamical structure of the boundary layer in urban areas. Proceedings from WMO Symposium on the interpretation of broadscale NWP products for local forecasting purposes. WMO - No. 450.

Speranskiy, L.S., V.N. Lykosov and L.N. Gutman, 1975: The turbulent planetary boundary layer above a curvilinear underlying surface in the presence of horizontal heat advection. *Izv. Atm. and Ocean. Phys.*, 11, (5), (English translation), 267-272.

Tapp, M.C. and P.W. White, 1976: A non-hydrostatic mesoscale model. *Quart. J.R. Meteor. Soc.*, 102, 277-296.

Taylor, P.A. and Y. Delage, 1971: A note on finite-difference schemes for the surface and planetary boundary layers. *Boundary-Layer Meteor.*, 2, 108-121.

Walmsley, J.L., 1973: Numerical experiments with an air-water boundary-layer model. *J. Meteor. Soc., Japan, Ser. 2*, 51, 388-399.

Welch, R. and W.G. Zdunkowski, 1976: A radiation model of the polluted atmospheric boundary layer. *J. Atmos. Sci.*, 33, 2170-2184.

Wu, S.S., 1965: A study of heat transfer coefficients in the lowest 400 meters of the atmosphere. *J. Geophys. Res.*, 70, 1801-1807.

Wyngaard, J.C. and O.R. Coté, 1974: The evolution of a convective planetary boundary layer - a higher order closure model. *Boundary-Layer Meteor.*, 7, 289-308.

Wyngaard, J.C., 1975: Modeling the planetary boundary layer-extension to the stable case. *Boundary-Layer Meteor.*, 9, 441-460.

Yamada, T. and G. Mellor, 1975: A simulation of the Wangara atmospheric boundary layer data. *J. Atmos. Sci.*, 32, 2309-2329.

Yamamoto, G. and A. Shimanuki, 1966: Turbulent transfer in diabatic conditions. *J. Meteor. Soc., Japan, Ser. 2*, 44, 301-307.

Yamamoto, G., A. Shimanuki, M. Aida and N. Yasuda, 1973: Diurnal variation of wind and temperature fields in the Ekman layer. *J. Meteor. Soc., Japan, Ser. 2*, 51, 377-387.

Yu, Tsann-Wang, 1976: Numerical studies of the atmospheric boundary layer with a turbulent energy closure scheme. AMS Third Symposium on Atmospheric Turbulence, Diffusion and Air Quality, Raleigh, N.C., October 1976.

Yu, Tsann-Wang, 1977: A comparative study on parameterization of vertical turbulent exchange processes. *Mon. Wea. Rev.*, 105, 57-66.

Zdunkowski, W.G. and F.G. Johnson, 1965: Infrared flux divergence calculations with newly constructed radiation tables. *J. Appl. Meteor.*, 4, 371-377.

Zdunkowski, W.G. and A.E. Barr, 1972: A radiative-conductive model for the prediction of radiation fog. *Boundary-Layer Meteor.*, 3, 152-177.

Zdunkowski, W.G. and D.C. Trask, 1971: Application of a radiative-conductive model to the simulation of nocturnal temperature changes over different soil types. *J. Appl. Meteor.*, 10, 937-948.

Zdunkowski, W.G., J. Paegle and J.P. Reilly, 1975: The effect of soil moisture upon atmospheric and soil temperature near the air-soil interface. *Arch. Met. Geoph. Bibl., Ser. A*, 24, 245-268.

CHAPTER 5 - OBSERVATIONAL METHODS AND PROGRAMMES

5.1 Observational Methods

5.1.1 General Considerations

Although extensive programmes using a variety of methods have been conducted over the past twenty or more years, our observational studies of the PBL are still incomplete and many important gaps in our knowledge exist. In many cases theoretical studies can give the general form for relationships but observational studies are needed to evaluate the constants. In general, to expand our knowledge of the PBL it is necessary that theoretical developments, model studies and experimental observations proceed together.

In this section some of the currently used observational systems will be described so that the reader can better understand the advantages and limitations of existing data sets and can better appreciate the possibilities for future observational programmes. As in other Chapters extensive use will be made of reference material. Some very relevant general references are: Atmospheric Technology Nos. 6 and 7; Schwerdtferger (1976); and Hess (1979).

The parameters of most interest to boundary-layer meteorologists are the mean and fluctuating values of wind, temperature, humidity, and, to a lesser extent, pressure. To measure the mean values the emphasis should be on sensors with a high degree of accuracy and able to provide a good average while for the fluctuation measurements the response time is critical. Generally the following criteria are relevant. The accuracy requirements for wind are $\pm$ 0.05 ms^{-1} and for temperature $\pm$ 0.01°C. The desirable accuracy for humidity would be an accuracy of 0.01 mb. The equivalent accuracy for wet-bulb sensors will vary with humidity (Polavarapu and Munn, 1967). However, this is, at present, usually not attainable so that an order of magnitude less in accuracy must be accepted.

The second consideration, after accuracy, should be the response time and averaging time. For the measurement of mean values the only restriction is that the response time must be small compared to the averaging period. Typically 1 minute or faster is acceptable. The averaging time to give the desired statistics varies with the parameter to be measured. Lumley and Panofsky (1964: see also Wyngaard, 1973) have shown that:

$$\tau \approx 2\,\tau_i\,\overline{(f^1)^2}/(a^2\,\overline{f}^2)$$

where τ is the averaging time required to determine the mean $\overline{f}$ to an accuracy a (e.g., for 1% accuracy, a = 0.01), $\overline{(f^1)^2}$ is the ensemble variance of f about its ensemble mean and τ_i is the integral scale of the variable f. The variable mean $\overline{f}$ could be the mean wind speed, in which case $f = U + u$ so that $\overline{f} = U$ and $\overline{(f^1)^2} = u^2$

or it could be higher order moments such as the skewness of temperature fluctuations or the flux of momentum ($\overline{wu}$).

For neutral conditions and 1% accuracy, the averaging time for the mean wind should be (to an order to magnitude):

$$T_U \sim \frac{10^3 z}{U} \text{ sec;}$$

that is, for U ~ 5 ms^{-1}, z ~ 5 m, T ~ 10^3 sec ~ 15 min; or for U ~ 5 ms^{-1}, z ~ 50 m, T ~ 10^4 sec ~ 150 min. Near the ground the sizes of the important eddies (represented by z/U) are relatively small and a good average can be obtained in 15 min. A similar argument would hold for temperature and humidity. For unstable conditions the eddy scale sizes would be larger but not likely by more than a factor of two. Therefore it would be acceptable to use averages of 15 min to 1 hr to get good estimates of the mean values of wind speed, temperature and humidity in the lowest 20 m or so of the PBL. However, for higher elevations it will be necessary to use longer averaging times, particularly in unstable stratifications. Unfortunately, the mean characteristics of the PBL are usually changing significantly over periods of more than an hour so that a meaningful average over time cannot be obtained. Hence, it is usually necessary to use an averaging period of about 1 hr and to recognize that the statistical reliability of the average decreases with height.

For the averages of turbulent fluctuations the requirements are more stringent. As shown by Wyngaard (1973), pp 136-141),

$$T_{\overline{w^2}} \sim T_{\overline{\theta^2}} \sim 4\, z/(a^2\, U)$$

for neutral and unstable conditions (z/L ~ -1), but for the fluxes of momentum ($\overline{uw}$) and heat ($\overline{w\theta}$) the averaging times are much larger, namely;

$$T_{\overline{uw}} \sim T_{\overline{w\theta}} \sim 20\, z/(a^2\, U) \quad \text{for near neutral conditions, and}$$

$$T_{\overline{uw}} \sim 100\, z/(a^2\, U)$$

$$T_{\overline{w\theta}} \sim 12\, z/(a^2\, U)$$

for unstable conditions. It is important to note that accurate flux estimates require several times longer averaging periods, than for the variances and for unstable conditions the required averaging time for momentum flux becomes very much greater than for heat flux. Practically one is forced to accept very much less statistical reliability in momentum flux values for unstable conditions or high elevations. This can be seen in plots of momentum flux estimates versus other parameters where the scatter of points will be very large.

The response time is an important consideration for measuring turbulent fluctuations which exist with time scales as small as milliseconds. Hot-wire anemometers (e.g., Friehe et al., 1975) can be used to resolve these microstructure variations with frequencies to a few k Hz. In most instances the concern is not with these very small scales but with the energy-containing eddies that are responsible for most of the turbulent variations and transfers. McBean (1972), Kaimal (1975) and others have shown that the range of significant contributions to the fluxes is approximately $0.001 < f < 3$ for near-neutral and unstable stratifications, where $f = nz/U$, the ratio of the wave length (n/U) to height, z. For stable stratifications there is a marked shift towards larger values of f (smaller scales) (Kaimal et al., 1972; McBean, 1972). For sufficiently stable stratifications ($z/L \sim 1$ or greater) the height is no longer the appropriate scaling parameter and should be replaced by L, the Monin-Obukhov length. In those cases the results indicate that eddies of scale size smaller than L/10 make no significant contribution to the flux. This corresponds to a high-frequency, small-scale cut-off of $nz/U \sim 50$, when $z/L \sim 5$.

In designing an experiment to measure the turbulent fluxes it is therefore necessary to ensure that the frequency response of the sensor is sufficiently greater than 3 U/z for most daytime conditions and as much as an order of magnitude faster for nocturnal conditions. Also, the sensor must have a spatial resolution of $(z/3)\ (1/2\pi)$ or better, where the 2π factor is included to avoid spectral attenuation. Thus a sensor with spatial resolution of 0.2 m (the value for typical sonic anemometers) should be used only at heights $z > 6\pi d \sim 4$ m or higher. For stable conditions the criteria would be $L > 20\pi d \sim 12.6$m.

In addition to ensuring the appropriate averaging and response times and accuracies for sensors it is necessary to have correct sensor exposure and mounting. Gill et al. (1967) have investigated the accuracy of wind measurements from towers and stacks. For an open triangular tower of equal sides D, the wake was found to have a width of about 1.5 D and extending downwind at least 6 D. Sensors mounted 2 D out from the corner of such a tower will usually measure speeds to within a $\pm$ 10% accuracy for an arc around the tower of 330°. Closed towers or stacks were found to give very much greater disturbance and it was necessary to mount sensors upwind 3 D from face of stack to get $\pm$ 10% accuracy. Izumi and Barad (1970) found for an open-lattice, rectangular tower that the wind speed error was as much as $\pm$ 5% at 1 D forward of tower. Wucknitz (1977) showed that the flow distortion depends on the distance in diameters (from a slim mast), on the angle of wind to the orientation of sensor-support arms and on Reynolds Number. Very large errors in flux estimates from the wind profiles were evident, particularly for asymmetrical mounting configurations.

For measurements of mean temperature sensors mounted in aspirated radiation shields are used. Although aspirated shields are necessary to reduce the radiation heating and to provide the ventilation for the sensor, the shields distort the flow and lead to errors in the representativeness of the measurement. Paulson et al. (1972) used unshielded but very small thermocouples to reduce radiation effects while allowing a natural flow around the sensor. For turbulence measurements it is, or course, essential not to seriously disturb the flow. An important additional provision pertains to the measurement of the vertical component of the wind, w. In order that the w measurement not be contaminated by horizontal wind variations, the sensors must be aligned to an accuracy of 0.1^{o}. Errors of as much as 25% per degree of tilt (Kaimal and Haugen, 1969) can occur in the momentum flux measurement.

The usual reason for micrometeorological measurements of mean wind, temperature and humidity is to determine their vertical gradients. Since the wind profile for neutral conditions is logarithmic the most common procedure is to place the sensors at logarithmic height intervals (i.e., 0.5, 1.0, 2.0, 4.0, 8.0 m). Generally, sensors should be placed close together in regions of large gradient and further apart in regions of small gradient. Hence, the sensor spacing should be designed with the project objectives clearly in mind. The derivatives can be approximated by:

$$\left.\frac{\partial F}{\partial z}\right|_{(z_1 z_2)^{1/2}} \approx \frac{F_2 - F_1}{(z_1 z_2)^{1/2} \ln (z_2/z_1)}$$

where F is the profile variable and z_1 and z_2 are the observations levels (Panofsky, 1965); Paulson, 1970). The approximation is valid for the geometric mean height of the two observation levels and is exact for a logarithmic profile. Paulson (1970) notes that the error is only 6% (and not significant compared to uncertainties in profile formulas) for $F \alpha z^{-1/3}$, $z_1 = 0.5$ and $z_2 = 16$ m.

Because the differences between adjacent levels are usually small the experimental program should include frequent intercalibration of sensors. This can be done by periodically positioning all sensors at the same level, by use of a well-controlled calibration facility or by moving a single sensor to each level. This latter procedure was followed by Paulson et al. (1972). Although the calibration problem is avoided serious sampling errors may be introduced. Donelan et al. (1974) has used fixed sensors at logarithmic intervals in height and a set of sensors on a trolley that moved up and down and, hence, provided a continuous calibration for the system.

Another important consideration in experimental design is to ensure that there is adequate fetch upwind of the tower. This is necessary to allow the turbulence and mean gradients to come into equilibrium with the surface. If the fetch is over surfaces of mixed characteristics then the measurements will represent, in some ill-determined way, a mixture of those surfaces. There is now a fairly extensive set of literature on the response of the surface layer to step changes in surface roughness and temperature (e.g., Rao et al., 1974; Taylor, 1969, 1970, 1971; Vugts and Businger, 1977; Lo, 1977). To a first approximation one can say that, at a distance x downwind of an abrupt change of surface roughness, the internal boundary layer to a depth $h_1 \sim x/100$ is in equilibrium with the new surface while above $h_2 \sim x/10$ the air is still in equilibrium with the old surface (the surface before the roughness change). Between those two heights the boundary layer is in a state of transition. Pasquill (1972) has discussed this in more detail and provided useful ratios for different stabilities and roughness conditions. The more general problem of many, rather than one, changes, in roughness and surface temperature and humidity, is still in need of further study.

5.1.2 Instruments: In-Situ Sensing

Instrument systems can be generally divided into two groupings: in-situ sensors that directly measure a property of the fluid in which they are immersed; and remote sensors that measure a property of the fluid at a position remote from where they are located. The in-situ sensors will be briefly reviewed in this section while the remote sensors will be the subject of Section 5.1.3.

Many types of sensors have been used for in-situ sensing of wind, temperature and humidity and it is possible here to only review the most commonly used. Kaimal (1975) has prepared an excellent quick summary of the advantages and limitations of various types of sensors and this is reproduced here as Tables 5.1 to 5.3. The following comments should be considered as additional clarification or information.

Cup anenometers are probably the most extensively used sensor in micrometeorological studies. The main errors occurring in cup anemometer measurements arise due to either overspeeding (a cup anemometer speeds up faster than it slows down) or shadows of supporting structures. The dynamics of cup anemometers have been extensively studied (e.g., Busch and Kristensen, 1977; Wyngaard et al., 1974, Kaganov and Yaglom, 1976) and are reasonably well understood. The overspeeding depends on anemometer design but typically amounts to a few percent (Hyson, 1972). Other errors can arise due to the influence of the vertical wind fluctuations and the imperfect cosine response of the cup anemometer (MacCready 1966) estimates this error as high as 6% at 4 m and 3.2% at 31 m for standard cups. Wind vanes, to give wind direction, have also been thoroughly investigated (e.g., Wieringa, 1967, 1972; Camp et al., 1970).

For measurements of wind turbulence sonic anemometers or propeller anemometers are the most used in atmospheric studies. Sonic anemometers are probably the best available sensor but are relatively expensive and require well-trained electronics personnel for maintenance. This has limited their use to relatively few research groups. Propeller anemometers, on the other hand, are relatively inexpensive and simple to operate. Their characteristics have been the subject of several recent papers to which the reader is referred (Hicks, 1972; Horst, 1973; McBean, 1972, 1975; Garratt, 1975; Wesley and Hicks, 1975). A common problem for any sensor used to measure the vertical component of the wind is to ensure that it is accurately aligned so that the horizontal wind fluctuations do not contaminate the w signal. This is particularly important for measurements of stress (the $\overline{uw}$ covariance). Kaimal and Haugen (1969) have shown that the error can be expressed as:

$$E = \frac{\sin 2\theta}{2}\left[\left(\frac{\sigma_u}{u_*}\right)^2 - \left(\frac{\sigma_w}{u_*}\right)^2\right]$$

where θ is the tilt angle. Dyer et al., 1970; (see also Kaimal and Haugen, 1971; Dyer et al., 1971) have suggested a way of reducing the error. However, the best

Table 5.1 - Sensors for Profile Measurement (reproduced from Kaimal, 1975)

Parameter	Sensor	Advantages	Limitations	Remarks
$\overline{u}$	Cup anemometer	With proper design, response essentially linear with speed above 1 m/s. Distance constant about 1 m on the average for lightweight low-friction types.	Starting speed 0.5 m/s, too high for light wind conditions. Measures speed, not $\overline{u}$, causing 2-3% overestimate. Overspeeding due to dynamic effects reported 5-10%.	Frequent calibrations and intercomparisons essential. Overestimate in speed can be determined $[\approx \frac{1}{2} \sigma^2(v)]$. Overspeeding can be measured and profiles corrected. Not recommended for light and variable wind conditions.
	Propeller anemometer on vane	Lightweight propellers exhibit lower starting speed and greater linearity than cups. Distance constant 0.4-1 m. Single calibration based on axial flow.	Off-axis response deviates from cosine law. Response limited by distance constant of vane, always larger than for propellers.	Frequent calibrations and intercomparisons essential. Not usable under light and variable wind conditions.
	Three-axis fixed propeller anemometer	Simple in operations. Not limited by vane response.	Errors due to deviation from cosine response significant.	Corrections for cosine response can be made to improve accuracy in wind profiles. Not recommended in applications where wind direction is highly variable.
	Two-axis sonic anemometer	No moving parts. Linear response down to 0.01 m/s. Absolute calibrations. Averaging length 0.25 m. Ability to resolve components, best among profile sensors. Particularly suitable for light and variable wind conditions.	Velocity defects 5-10% observed on axis when wind direction parallels it. Error highest for steady wind direction.	Because it is an absolute device, its calibration is fixed by design parameters. Periodic checking of electronic adjustments necessary to insure maximum accuracy. Used successfully in AFCRL's 1973 Minnesota experiment. See Fig. 1.

Table 5.1 - Sensors for Profile Measurement - (Continued)

Parameter	Sensor	Advantages	Limitations	Remarks
T	Resistance thermometers	Low cost, simple to construct. Particularly suited for temperature difference (ΔT) measurement. ΔT accuracy of $0.01^{\circ}C$ possible with great care in design of probes and radiation shields. Time constant depends on aspiration, sensor mass, specific heat, etc., ~ 1 min.	Nonlinearity in calibration curve and cumulative errors limit accuracy in absolute temperature measurement to $0.1^{\circ}C$. Calibration for ΔT is a function of T.	ΔT measurements between successive levels plus one between top and bottom to serve as a check; one T measurement to serve as reference located at any convenient observation level.
	Thermocouples	Same as above except smaller time constant possible with small junctions.	Output roughly an order of magnitude smaller compared to resistance bridges. Stray emf's from lead junctions to be avoided. Best accuracy $0.02^{\circ}C$.	Same as above.
	Quartz crystal thermometer	Linearity, accuracy, and long-term stability, better than $0.01^{\circ}C$ for T measurement over range -20 to $+40^{\circ}C$.	Relatively expensive. Output is digital form, requiring D/A converters or buffers for digital data transfer. Marketed by only one manufacturer.	Used for profile measurements in AFCRL's 1973 Minnesota experiments with aspirated radiation shield of silvered vacuum flask construction (commercially available, modified to accept sensor). See Fig. 1.

Table 5.1 - Sensors for Profile Measurement - (Continued)

Parameter	Sensor	Advantages	Limitations	Remarks
$\overline{q}$	Wet-and dry bulb psychrometer	Any temperature sensor above can be adapted. With great effort, 2% accuracy in relative humidity can be attained. Time constant 1-2 min.	Psychrometric constant function of ventilation at low wind speeds. Accuracy depends on design of wick and purity of water used for wetting. Accuracy decreases with decreasing relative humidity and temperature.	Conversion of wet-bulk depression to vapor pressure and specific humidity require use of psychrometric tables. Frequent change of wicks and use of distilled water essential for accuracy.
	Dew-point hygrometer	More fundamental method than wet-bulb depression method. Conversion of dew-point to vapor pressure not based on empirical constants as in wet-bulb method. Accuracy $\sim 0.25^{o}C$.	Errors increase with decreasing dew-point temperature.	Commercial versions available with Peltier cooling of mirror and photoelectric detection of mirror surface. Temperature sensor is embedded in the mirror.

Table 5.2 - Sensors for Flux Measurement (reproduced from Kaimal, 1975)

Parameter	Sensor	Advantages	Limitations	Remarks
u', v', w'	Hot-wire and hot-film anemometers	Frequency response extends to several kHz. Small size. Ideal for use in first 4 m above ground. Various configurations of two- (and three-) wire probes for sensing flow direction fluctuations in two (and three) dimensions.	Calibration drifts with ambient temperature, aging, and contamination of wire. Somewhat fragile. Multi-wire probes which measure angular fluctuations must be oriented into the wind. Large wind direction shifts not followed accurately.	Successful use of hot-wire techniques was made by the Danish AEC during the 1968 Kansas experiments. Probes were attached to a rapid-response vane and replaced frequently before they had a chance to break or become contaminated. Each probe was calibrated (before and after the experiments) over the range of ambient temperatures encountered.
	Three-axis sonic anemometer	Frequency response limited by path length (20 cm). Sensitive only to wind components along the acoustic paths. Calibration very stable.	Path-averaging limits its use to heights 4 m and above. Probe should be oriented approximately into the wind to avoid velocity defect along paths. Acceptance angle ($\pm 30^{\circ}$), larger than for hot wires.	w is measured directly, u and v are derived from two horizontal axes, 120° apart. Probes should be mounted on antenna rotor and controlled remotely to face anticipated wind direction. Used in AFCRL's Kansas and Minnesota experiments. See Fig. 2.
	Anemoclinometer	Small size. Wind components determined from pressure readings at a number of points on a sphere.	Probe needs to be oriented into the wind. Frequency tesponse ($\sim$ 10 Hz) limited by time constant of pressure transducers.	Despite small size, fluxes for heights below 1 m will be underestimated because of poor frequency response.

Table 5.2 - Sensors for Flux Measurement - (Continued)

Parameter	Sensor	Advantages	Limitations	Remarks
	Three-axis propeller anemometer	Simple, low cost.	Distance constant is a fraction of angle of attack: ∿ 0.4-1.4 m for angles 0-80° from axis. Deviations from cosine response cannot be ignored; most serious (∿ 30%) for angles close to 90° from axis.	Propeller sensing w should be calibrated for angles of attack 60-120° from axis. Distance constant for w may be too large for accurate flux measurements below 20 m. (Fixed propeller for w used with propeller on vane for horizontal speed found better for stress measurement. Error from using speed rather than u less serious than error from cosine response.)
	Propeller on bivane	Simple, moderate cost.	Response limited by distance constant of vane, 1 m at best.	In general not suitable for flux measurements below 20 m.
T'	Platinum wire thermometer	Exposed lengths of platinum wire (∿ 25 μm diameter) wound or strung on supports designed to mate with existing anemometer.	Frequency response on the order of 10-20 Hz for typical wind speeds in the surface layer.	Temperature probe designed to fit within the frame of a three-axis sonic anemometer used in Kansas and Minnesota experiments. See Fig. 2.
	Hot-wire probe in constant-current mode	Same principle as above but has faster response (300-400 Hz) and better spatial resolution. Hot-wire anemometer electronics can be adapted with only simple modifications.	Wire very thin (5 μm diameter), fragile.	Ideal for use with hot-wire single- and X-probes for heat flux measurements close to the ground.

Table 5.2 - Sensors for Flux Measurement - (Continued)

Parameter	Sensor	Advantages	Limitations	Remarks
	Sonic thermometer	Measurement made along same path as w. No extra hardware needed. Response limited only by path length.	Measured temperature actually approximates virtual temperature. Also severe contamination of signal by wind component normal to path observed when operated in near-neutral and stable conditions.	Suited only to measurements in dry air under convective conditions with strong temperature fluctuations and light winds.
q'	Lyman-alpha humidiometer	Rapid response. Spatial resolution 5-20 mm typical. Device relatively uncomplicated. Measures absolute humidity.	Calibration drifts due to changes in source and detector characteristics, dirt on window and deterioration of window from moisture in the air.	Response curve is approximately logarithmic, but slope of linearized output stays reasonably constant despite shifts in absolute position. Mechanical configuration of sensor interferes with full exposure to wind, but preferable to aspiration through intake tube.
	Infrared hygrometer	Open transmission path (25-50 cm) provides less interference to flow. Response limited by path averaging and time constant of the mechanical components ($\sim$ 10 Hz). Measures absolute humidity.	Complex optical mechanical system involved. Calibration stability and noise from mechanical components are serious problems.	Open-path design, in which reference signal is provided by nonabsorbing frequency along same optical path, is inherently less stable than closed-path systems where reference signal is the same frequency through parallel path in dry air. The latter technique required aspiration through intake tube, not suitable for eddy flux measurements.

Table 5.2 - Sensors for Flux Measurement - (Continued)

Parameter	Sensor	Advantages	Limitations	Remarks
	Thin-film hygrometer	System relatively simple (typically film resistance or capacitance measurement). A 10 Hz response is attainable; fine spatial resolution (∿ 1 cm) is possible.	Calibration curve varies with production technique. Since it is a bulk absorption device, it responds to relative humidity.	If substrate temperature differs from ambient temperature (due to radiation), readings will be altered. Substrate temperature should be monitored for best accuracy. Full potential of this technique yet to be realized.

Table 5.3 - Sensors for Measuring Fluxes at the Earth's Surface (reproduced from Kaimal, 1975)

Parameter	Sensor	Advantages	Limitations	Remarks
τ	Shear-stress meter	Measures τ directly. The averaging period required for stable estimates $\sim$ 15 min, less than for eddy-correlation method.	Matching of sample tray to surroundings critical. Calibration requires verification from eddy-correlation measurements close by.	Sample tray used in Minnesota experiments is 2 m in diameter (shown with cover on Fig. 1). Frequency response 1 Hz.
E	Lysimeter	Directly measures water lost by evaporation from surface.	Simulation of true moisture gradient in the soil not easy. Systems tend to be massive ($\sim$ 6 m diam). Not easily transportable.	Permanent test site needed for use of this technique.

procedure seems to be to reduce the tilt angle (deviation from vertical) to the order of 0.1° if possible. Over land surfaces the additional problem arises in that the surface may have a mean slope of more than 0.1° and, in that case, it is not entirely clear whether one should align with gravity or with the local perpendicular to the surface.

Other sensors that have been used to measure the turbulent wind, pressure or fluxes are described in articles by Friehe et al. (1975), Dyer (1975), Elliott (1975) and Yap et al. (1974). Measurement of temperature fluctuations is relatively more straightforward, using resistance wire or thermistor thermometer. Very fine wires (0.25 μm diameter platinum) have been used by Boston and Burling (1972) while more robust longer path-length wires were employed by Kaimal (1968). Humidity fluctuation measurements are more difficult. Among the techniques used are: absorption of Lyman-α radiation (Miyake and McBean, 1970; Buck, 1973); dew-point hydrometer (Miyake and McBean, 1970); microwave refractometer (Bean and Gilmer, 1969); wet and dry-bulb thermocouples (Polavarapu, 1972); and infrared humidometer (Yelagina, 1969).

Although most boundary-layer studies have used towers to support sensors there has been an increasingly important use made of aircraft and balloons. Aircraft have several advantages; principally, they can fly at specified levels above about 30 m and they can provide spatial averages. Their disadvantages are primarily the cost and complexity of the systems required. In order to measure winds from an aircraft it is necessary to first measure the air speed relative to the aircraft and then measure the aircraft speed relative to the ground. The former can be measured with a pitot tube and vanes (e.g., Lenschow, 1971, 1975) or sonic anemometer and hot-wire anemometer (Miyake et al., 1970). The pitot tube is the most used device obtaining mean along axis air speed while the vanes or sonic/hot-wire anemometers are used to obtain measurements of the turbulent fluctuations. The turbulence measurements must be made well away from the aircraft fusilage to avoid flow distortion. Usually a boom forward of the nose is employed but some experimenters (e.g., Yokoyama et al., 1973; Sheih et al., 1971) have mounted their sensors on wing tips or wing struts. Measurement of aircraft motions with respect to the ground are usually more difficult and expensive. Two main approaches are in use. The best and most expensive is the use of inertial navigation systems (INS) (e.g., Axford, 1968; Lilly and Lenschow, 1971). An alternate approach is to utilize accelerometers and gyroscopes to measure aircraft orientations and motions (e.g., Miyake et al., 1970; McBean and MacPherson, 1976). This type of system can measure the turbulent velocities almost as well as those using inertial navigation systems but the measurements of mean wind are considerably less accurate due to drift in accelerometers. For mean wind measurements several aircraft systems employ doppler radars to measure aircraft ground speed. For complete description of such a system see MacPherson (1973, 1974). Some information can be obtained just from aircraft response (Lenschow, 1976).

Temperature is usually measured with a resistance wire or thermistor. Because of the aircraft speed the exposed probe will be dynamically heated. To reduce this effect special probe designs are necessary. The correction for converting the measured temperatures to the ambient temperature is proportional to the square of the true airspeed and is about 5 °C at 100 m/s (Lenschow, 1975).

Instruments for measurement of humidity fluctuations include the Lyman-α humidiometer (Miyake et al., 1970) and microwave cavity refractometer (Friedman et al., 1970). Measurements in clouds are particularly difficult (Lenschow and Pennel, 1974).

The use of aircraft in boundary layer experiments is increasing and most major experiments have included them. A partial listing of some aircraft experiments and papers derived from them follows:

Experiment	Reference
- fluxes over Atlantic Ocean	Bunker (1955, 1960)
- BOMEX	Donelan and Miyake (1973), Bean et al., (1972), Grossman and Bean (1973)
- IFYGL	McBean and Paterson (1975), Bean et al., (1975)
- Trade-winds area	Pennell and Lemone (1974), Lemone and Pennell (1976).
- GATE	Merceret (1976), GATE Reports 11, 14, 18 (available from WMO)
- AMTEX	Lenschow and Agee (1976), Jensen and Lenschow (1978) Wyngaard et al., (1978).
- boundary layer modification	Lenschow (1973).
- over South Pacific	Warner (1971, 1972).

For measurements up to 1 km elevation a tethered balloon system is a good choice. Tethered balloon systems have the advantages of being relatively inexpensive (compared to aircraft or tall towers) and relatively mobile (compared to towers), of being used either for profile measurements or fixed at one level and of being able to accommodate a variety of instrument types, provided they are light enough to be lifted. Their disadvantages are that they are limited to relatively good weather (no high winds or electrical storms), the balloon motions may effect the measurements and the systems may be a hazard to aircraft so that air traffic control authorities may restrict or forbid its use. A tethered balloon system consists of five basic components: 1) a balloon, usually aerodynamically shaped and filled with helium; 2) a tether line; 3) a winch system to raise or lower the balloon; 4) instrument packages that can be suspended directly from the balloon; or fastened to the tether line; and 5) a telemetry and recording system.

Tethered ballon systems of several types now exist. Their characteristics have usually been determined by the scientific interests and backgrounds of the groups that developed them. The system developed in the United Kingdom (Readings and Butler, 1972) has been in use for about twenty years and is used primarily to measure turbulence. The system (Readings, 1975) incorporates a large barrage balloon (1300 m^3), steel cable and an instrument package to measure wind and temperature fluctuations. The wind sensors are "double V" hot-wire anenometers which are gymbal mounted on a pendulum arm and an eight-cup anenometer. The system has been used to investigate high frequency velocity (Rayment, 1973) and temperature (Caughey and Rayment, 1974) fluctuations, turbulence over the sea (Thompson, 1972) and structure of an inversion (Rayment and Readings, 1974) and its measurements have been compared with tower measurements (Haugen et al., 1975). Recently, the system was used to extend tower measurements up through the PBL (Kaimal et al., 1976; Caughey and Kaimal, 1977).

In the USA several tethered-balloon systems are in use. A small portable system has been developed by the National Center for Atmospheric Research. It can be operated by one man and senses pressure, temperature, wet-bulb temperature, wind speed and direction (Morris et al., 1975; Deloach et al., 1976). This system was used in the International Field Year on the Great Lakes to investigate lake breezes (Estoque et al., 1976). A larger system was used on the U.S. GATE ships. It had the capability of carrying several instrument packages and was called the Boundary-Layer Instrumentation System (BLIS) (Burns, 1974/5).

During GATE tethered balloons were used on ships from the U.K., Federal Republic of Germany, U.S.S.R. and Canada as well as the U.S.A. (See GATE Report No. 5). The Canadian system and the results of using the same system on two ships are described by Mickle and Davison (1979). A gymbal-mounted pendulum was used to keep minature cup and blade anemometers orientated to measure horizontal and vertical wind speeds. A French system using a hot-wire anemometer to measure turbulence is described by Weill et al. (1976). A Japanese system has been used in both urban areas (Yokayama, 1969) and over the sea in AMTEX.

5.1.3 Instruments - Remote Sensing

Remote sensing of the boundary layer involves the use of radio, optical and acoustic techniques to measure winds, temperatures or atmospheric conditions along a line, over an area or throughout a volume. The field is one of the most rapidly developing in meteorology and Hooke (1978) has described it as being "currently at the crossroads" with working prototypes of most important instruments in existence and new applications being sought. When acoustic, radio, and optical waves propogate through the atmosphere they interact with the atmosphere and their changes depend on atmospheric properties. Three examples are: the scattering of acoustic waves depends on the temperature and wind turbulent fluctuations of the medium; the doppler shift of radio waves depends on the velocity of the turbulent eddies relative to the waves; and the absorption of an optical wave depends on the composition of the gaseous mixture through which it is propagating. Remote sensors can work in both the passive (sensing naturally occurring radiation) and active (producing their own radiation) modes. For boundary-layer studies, just about all systems in use are active ones, which have the advantage of improved spatial resolution, flexibility and scanning capability. The comparative advantages of radar, sodar and lidar are discussed by Browning et al. (1973).

The advantages of remote sensors are primarily their ability to give good spatial and temporal resolution, representative averages and to measure without placing a sensor in the field. In addition, although the systems are initially relatively expensive, they can produce an abundance of information, so much so that the incremental cost of each observation is small.

Although acoustic wave propagation can be used in a variety of ways to sense the atmospheric properties, the most frequent use is in the form of acoustic echo sounding or acoustic radars (sometimes called sodars). Recently Brown and Hall (1978) have reviewed the field. Earlier reviews by Morris and Hall (1974-75), Beran (1974-75), Little (1972) and McAllister et al. (1969) are also useful sources of information. The acoustic radars were originally operated to receive the scattered and returned echo from turbulent inhomogeneities in the boundary layer and are now used in the doppler mode as well to sense the turbulent velocities. The scattering of sound by velocity and temperature fluctuations can be written in terms of the inertial subrange scattering cross-section, $\sigma(\alpha)$ (Morris and Hall, 1974-75):

$$\sigma(\alpha) = 0.05\ \lambda^{-1/3} \cos^2\sigma \left[\frac{C_v^{\ 2}}{c^2} \cos^2\frac{\sigma}{2} + 0.13\ \frac{C_\theta^{\ 2}}{T^2}\right] \left(\sin\frac{\alpha}{2}\right)^{-11/3}$$

where: α, is the scattering angle measured from the direction of propagation;

λ is the wavelength of the sound for average temperature T (assumed that the turbulence causing the scattering is locally isotropic),

c is the speed of sound in the undisturbed atmosphere,

C_v and C_θ are the velocity and thermal structure functions defined as:

$$C_v^{\ 2} = \overline{\left[\frac{v(x) - v(x+\Delta x)}{(\Delta x)^{1/3}}\right]^2}$$

$$C_\theta^{\ 2} = \overline{\left[\frac{\theta(x) - \theta(x+\Delta x)}{(\Delta x)^{1/3}}\right]^2}$$

It will be noted that the scattering is weakly dependent on wavelength with shorter waves being scattered more strongly and is highly dependent on scattering angle, α.

For backscatter (the original and still most common use), $\alpha = \pi$, the equation reduces to:

$$\sigma(\alpha) = 0.055\ \lambda^{-1/3} \left[0.13\ \frac{C_\theta^{\ 2}}{T^2}\right]$$

and, hence, backscattered sound is dependent on the temperature fluctuations and not the velocity fluctuations. On the assumption that there is temperature fluctuations throughout the mixed layer, acoustic radars are frequently used to delineate the depth of the mixed layer. For convective boundary layers the concepts of similarity theory can be applied to relate the value of C_θ^2 (and hence the scattering cross section) to the surface heat flux, Q_o (Wyngaard et al., 1974):

$$C_\theta^2 \simeq (4/3\ k^{2/3})\ (T/g)^{2/3}\ (Q_o/z)^{4/3}$$

If a receiver is displaced from the transmitter (the bistatic mode) then information on C_v^2 can be obtained. Kerman (1976) has presented further information on the interpretation of acoustic sounder returns.

The doppler shift of the returned acoustic echo can provide information on wind velocity. Kaimal and Haugen (1977) and Hardesty et al. (1977) have described two possible systems. The additional doppler data plus the information on C_θ and C_v can be used to provide considerable quantitative information on the boundary layer, including profiles of wind and turbulence and the surface fluxes of heat and momentum (Brown and Hall, 1978).

Radar techniques for probing the boundary layer have been of two types: pulse doppler radar (e.g., Browning et al., 1973a; Lemon et al., 1977) and FM-CW radars (e.g., Bean et al., 1973; Chadwick et al., 1976 a,b). The pulse doppler radars have proven to be a powerful tool for investigating the wind patterns over urban areas using chaff dispersed from aircraft as tracers (e.g., Kropfli and Kohn, 1977). Better than 1 km resolution over areas greater than 100 km^2 can be obtained (Hooke, 1978). Thus pulse doppler radars could be used for initialization and verification of mesoscale models. Readings et al., (1973) have used a vertically pointing doppler radar to examine the characteristics of an inversion on top of a convective boundary layer. FM-CW radars have also been used to investigate wave motions in stable layers (Gossard et al., 1973). The major limitation of FM-CW radars is that their range is limited to a few kilometres as opposed to a few tens of kilometres for pulse doppler radars but they do provide better spatial resolution (Browning et al., 1973b).

Laser radars or lidars are the most popular optical technique. Huffaker (1970, 1974-75) describes the principles of doppler lidar velocity determination. Other wind studies using lidars have been described by Benedetti-Michelangeli et al., (1972) and Schwiesow and Cupp (1976). Some success can also be expected for the measurement of temperature profiles from the molecular thermal broadening (Fiocco et al., 1971), from Raman-scatter lidar estimates of N_2 scale heights (Strauch et al., 1971) or from rotational Raman scattering (Cooney, 1977). Lidars also have the potential for measuring water vapour (Strauch et al., 1972) and gaseous pollutants. This area is at present rapidly developing. Other optical techniques for remote probing include line-of-sight observations which have been used to measure winds (Lawrence et al., 1972), surface convergence (Kjelaas and Ochs, 1974) and surface heat flux (Wesley, 1976; Wyngaard et al., 1978).

Most recently developments have turned to the use of so called "hybrid" systems that are combinations of techniques for probing the boundary layer. One example is the work of Noonkester et al., (1974) comparing FM-CW radar and lidar returns. Another approach is to combine radio and acoustics techniques in the RASS (radio-acoustic sounding system). In this system temperature profiles are measured by using radiowave pulses to follow an acoustic pulse up through the atmosphere (Frankel et al., 1977).

Although this brief review has only mentioned a few of the techniques and new developments it will be appreciated that remote sensing of the boundary layer is a rapidly growing and important field. Remote sensing methods and instruments can now provide information in three dimensions and time on winds, turbulence, humidity, temperature and concentrations of certain pollutants. Hopefully their use will become more widespread as operational units requiring less technical expertise to operate and of low cost become available.

5.2 Operational Programmes

Over the past few decades there have been a number of boundary-layer observational programmes. In some cases these have been organized specifically for boundary-layer studies while in other cases the boundary layer studies have formed a component, usually an important and major component, of larger experiments. Each of these observational programmes have lead to a furthering in our understanding of boundary-layer processes and have provided data sets that are invaluable to the field. In many cases these data sets have or can be used by other scientists to test hypotheses or models or for further analysis. In the previous section the types of observational methods have been described. In this section a tabulation of boundary-layer observational programmes is given so as to acquaint interested scientists with the extent of these programmes. Four basic types of information are given for each programme; namely:

1. a brief description of the programme including its objectives;
2. a listing of the type of data collected;
3. one or more scientific papers that have been published that give some further information on the observational programme; and
4. an information source for further information and data, if possible.

It is anticipated that this information will be useful both for scientists considering the design of boundary layer experiments, in that they can benefit from what has gone before (and for perhaps deciding that another experiment is not necessary) and for scientists needing data to further their scientific activities.

A. AFCRL Kansas 1968 Field Program

1. The Boundary Layer Branch of the United States Air Force Cambridge Research Laboratories conducted a field experiment in July and

August, 1968 at a site in southwest Kansas. The objectives were to make direct measurements of vertical transport of momentum and sensible heat, as well as vertical profiles of wind and temperature in the atmospheric surface layer over a horizontally uniform, flat terrain.

2. Turbulent fluxes, profiles.

3. Businger, J.A., J.C. Wyngaard, Y. Izumi and E.F. Bradley, 1971: Flux-profile relationships in the atmospheric surface layer. J. Atmos. Sci., 28, 181-189.

4. Kansas 1968 Field Program Data Report, Editor - Y. Izumi, AFCRL-72-0041, Environmental Research Papers No. 379, Air Force Cambridge Research Laboratories (YLB), Hanscom AFC, Massachusetts 01731, U.S.A.

B. AFCRL Minnesota 1973 Atmospheric Boundary Layer Experiment

1. The primary objectives of the Minnesota Experiment were to make measurements of the vertical fluxes of momentum and heat and of profiles of wind velocity and temperature within the planetary boundary layer.

2. Fluxes and profiles: tower measurements to 32 m and tethered balloon measurements to 1220 m.

3. Kaimal, J.C., J.C. Wyngaard, D.A. Haugen, O.R. Cote, Y. Izumi, J.S. Caughey and C.J. Reading, 1976: Turbulence Structure in the Convective Boundary Layer. J. Atmos. Sci., 33, 2152-2169.

4. Minnesota 1973 Atmospheric Boundary Layer Experiment Data Report, Editors - Y. Izumi and J.S. Caughey, AFCRL-TR-76-0038, Environmental Research Papers, No. 547, Air Force Cambridge Research Laboratories (LYB), Hanscom AFB, Massachusetts 01731, U.S.A.

C. AIDJEX - Artic Ice Dynamics Joint Experiment

1. AIDJEX was a study of ice dynamics and atmospheric-oceanic forcing in the artic ice pack. Pilot projects were conducted in 1970, 1971 and 1972 while the main field program was in 1975-76.

2. Boundary layer data include air pressure, inversion height and some profile and turbulence information.

3. AIDJEX Bulletins - AIDJEX Data Manager, 4059 Roosevelt Way N.E., Seattle, Washington 98105, U.S.A.

D. AMTEX - Air Mass Transformation Experiment

1. AMTEX was an experiment to clarify the transfer processes by which energy and momentum are supplied from sea surface to the air and transported to the free atmosphere through the p.b.l., especially in areas where the air-mass transformation is active. Experiments conducted February-March, 1974 (AMTEX, 74) and February-March, 1975 (AMTEX, 75).

2. Boundary layer data were collected from towers, tethered balloons, ships and aircraft.

3. Scientific Reports of 1st, 2nd, 3rd, 4th AMTEX Study Conferences - available from Prof. Y. Mitsuta, Disaster Prevention Research Institute, Kyoto University, Uji, Kyoto, 611, Japan. AMTEX - GARP Publication No. 13 - available from WMO.

4. Data Reports - AMTEX 74 and AMTEX 75 available from: University of Tokyo Press, 7-3-1 Hongo, Bunkyo-ku, Tokyo 113-91, Japan.

E. ATEX - Atlantic Tradewind Experiment

1. The purpose of ATEX was to investigate comprehensively the energy, momentum and water vapour balances in the Tradewind region. The experiment was conducted February, 1969 and utilized four ships in the tropical Atlantic Ocean.

2. Ocean flux and profile measurements from a buoy, radiosondes from ships.

3. Dunckel, M., et al., 1974: Turbulent fluxes of momentum, heat and water vapour in the atmospheric surface layer at sea during ATEX. Boundary-Layer Meteor., 6, 81-106.

4. Augstein, E., et al., 1974: The vertical structure of the atmospheric boundary layer in undistrubed Tradewinds over the Atlantic Ocean. Boundary-Layer Meteor., 6, 129-150.

F. BOMEX - Barbados Oceanographic and Meteorological Experiment

1. BOMEX was a large-area air-sea interaction experiment conducted in the Atlantic Ocean area east of Barbados during the period May 3 and July 11, 1969.

2. Boundary layer data were collected with ships' booms, tethered balloons, aircraft, buoys and other platforms.

3. Holland, J.Z., R.W. Reeves and W.Y. Chen, 1976: An assessment of BOMEX Scientific Results. NOAA Tech. Report.

4. Centre for Experimental Design and Data Analysis, NOAA Environmental Data Service, Page Building 2, Washington, D.C. 20235, U.S.A.

G. CAENEX - The Complete Atmospheric Energetics Experiment

1. The general purposes of CAENEX are: the investigation of the transfer of all categories of energy and all kinds of heat flux divergence in the atmosphere; the analysis of the factors which are responsible for these processes under various conditions; the elaboration, on this basis, or recommendations for taking account of energetic factors of the thermal regime and dynamics of the atmosphere. As such the program has as one of its sub-objectives the investigation of boundary layer energy fluxes and flux divergences. The program is an on-going contribution to GARP being undertaken in the USSR.

2. Radiation and boundary layer fluxes using profiles, eddy flux measurements and aircraft.

3. CAENEX - GARP Publication No. 12 - available from WMO (which describes the experiment and gives the results CAENEX 70, 71, 72).

H. EKAM 73 - Interaction of the Sea and the Atmosphere in the Nearshore Zone

1. This program was carried out between May 15 and June 30, 1973 in the southern Baltic Coast with the principal objective of studying the processes of formation and transformation and of the structure of certain oceanological and meteorlogical fields in the near-shore zone of the sea up to a distance of about 300 m from the shore. (EKAM = Einflüsse der Küste auf Atmosphäre und Meer).

2. Turbulent fluxes, microstructure, profiles, waves, currents.

3. The Interaction of the Sea and the Atmosphere in the Nearshore Zone. Proceedings of the EKAM 73 international field study in the Baltic Sea Nearshore Zone. Raporty, Seria R, NR la, Gdynia 1975, Morski Instytut Rybacki, Osrodek Wyndawniczy, Poland.

I. GATE - GARP Atlantic Tropical Experiment

1. The Boundary-Layer Sub-Programme (BLSP) of GATE was one of five sub-programmes. The BLSP objectives were: the testing of existing b.l. models and parameterization schemes, provision of an observational basis for the development of improved models and to support further advancements and developments. The experiment was conducted in the tropical Atlantic ocean region in June-September, 1974.

2. Ships, tethered balloons, aircraft.

3. GATE Report Series - available from the World Meteorological Organization.

   No. 5 - Boundary Layer Sub-Programme; No. 14 - Preliminary Scientific Results, (Vols. I and II); No. 19 - Report of Field Phase - Summary of Data Collected; No. 20 - The Final Plan for the GATE Sub-Programme Data Centres.

4. BLSP Data Centre - Max Plank Institut fur Meteorologie, Bundestrasse 55, 2 Hamburg 13, Federal Republic of Germany.

J. Great Plains Turbulent Field Program

1. The Great Plains Turbulence Field Program was conducted near O'Neill, Nebraska in the summer of 1953. The objectives were to investigate the atmospheric boundary layer over a flat prairie surface.

2. Turbulence, profiles, energy budget components.

3. Lettau, H.H. and B. Davidson, 1957: Exploring the Atmosphere's First Mile. Pergamon Press Inc., N.Y.

K. IFYGL - International Field Year on the Great Lakes

1. IFYGL was a comprehensive study of the physical processes related to Lake Ontario and the basin during the period April 1, 1972 to March 31, 1973.

2. Boundary layer data were collected from buoy and shoreline networks, aircraft, towers and other platforms.

3. Elder, F.C., J.Z. Holland and J.A. Almazan, 1974: IFYGL Atmospheric Boundary Layer Program - Summary and Status of Results. Proc. IFYGL Symp., 55th Ann. Mtg., AGU, NOAA, U.S. Dept. of Commerce, Rockville, Maryland, pp. 70-85.

4. U.S. IFYGL Project Office, Great Lakes Environmental Research Laboratory, 2300 Washtenaw Avenue, Ann Arbor, Michigan 48104, U.S.A., or Canadian IFYGL Centre - CCAH, Atmospheric Environment Service, 4905 Dufferin Street, Downsview, Ontario, M3H 5T4, Canada.

L. IMST Air-Sea Interaction Simulation Facility

1. The Institut de Mecanique Statistique de la Turbulence, Marseille, France, has designed and constructed a large wind-water tunnel allowing a partial simulation in the laboratory of the natural air-sea interaction processes.

2. Measurements of velocity, temperature and humidity (means and turbulence) and waves are possible.

3. Coantic, M. and A. Favre, 1973: Activities in and preliminary results of air-sea interaction research at I.M.S.T. Second IUTRAM-IUGG Symposium on Turbulent Diffusion in Environmental Pollution, 8-14 April, 1973, Charlottesville, Virginia. Adv. in Geophys, Vol. 18.

4. Dr. M. Coantic, Institut de Mechanique Statistique de la Turbulence 12 Avenue du Général Leclerc, 13003, Marseille, France.

M. Intercomparisons of Boundary Layer Instrumentation

1. To date three main intercomparisons of boundary layer instrumentation have been held. The first was held in the summer of 1968 at an over water test site of the Institute of Oceanography of the University of British Columbia, Vancouver, B.C., Canada. The second was held in 1970 at the Tsimlyansk field station of the Institute of Atmospheric Physics, Academy of Sciences of the USSR. The third was held in September, 1976 at the field site of the Division of Atmospheric Physics, CSIRO, Australia.

2. Eddy flux and profile measurements by a variety of instruments.

3. Vancouver: Miyake, M., et al., 1970: Comparison of turbulent fluxes over water determined by profile and eddy correlation techniques. Quart. J. Roy. Meteorol. Soc., 96, 132-137; Vancouver: Miyake, M., et al., 1971: Comparison of acoustic instruments in an atmospheric turbulent flow over water. Boundary-Layer Meteor., 2, 228-245; Tsimlyansk: Tsvang, L.R., et al., 1973: Comparison of turbulent measurements by different instruments. Tsimlyansk Field Experiment, 1970. Boundary-Layer Meteor., 3, 449-521; Australia: Contact Dr. A.J. Dyer, CSIRO.

N. JASIN - Joint Air Sea Interaction Project

1. The objective of JASIN is to understand the physics of small scale mixing and transfer processes in oceanic and atmospheric boundary layers. Measurements of turbulent structure of velocity, temperature and salinity in the mixed layer are a post objective. JASIN 72 was 1 month in 1976 and JASIN 77 is planned for 6-8 weeks in 1977 in the Atlantic to the west of Scotland.

2. Radiosondes; atmospheric flux measurements from balloons, aircraft and fixed stations; SST from aircraft; STD's and current in the upper layers of the ocean; surface waves.

3. Reports of 8, 9 and 11th Sessions of J.O.C. of GARP available from the WMO.

4. Royal Society Air-Sea Interaction Project, The Royal Society, 6 Carlton House Terrace, London SW1Y 5AG, England.

O. JONSWAP - Joint North Sea Wave Project

1. The JONSWAP objective is to understand the mechanisms of the transfer across the air-sea interface and the dependence of this transfer on sea-state and the energy balance of surface waves. A pilot study was held in September, 1968; the first experiment, July 1969; second experiment, September, 1973; third experiment; July-September, 1975; and wave-follower experiment, July-September, 1976.

2. Six-eleven wave stations, flux measurements, wave-follower measurements, active and passive microwave measurements from aircraft.

3. JONSWAP Report, December 1974. Available from: Max-Planck Institut für Meteorologie, 2 Hamburg 13, Bundesstrasse 55, Federal Republic of Germany. Hasselmann, K., et al., 1973: Deutsche Hydrogr. Zeitscher., Ergaenzungsheft Reihe A (8°), No. 12.

P. NORPAX - North Pacific Experiment

1. The objective of NORPAX is to study the North Pacific fluctuation on scales of months to years and greater than 1000 kms and their relation to atmosphere. Studies of SST and other data, mixed layer dynamics, special field studies and numerical modelling constitute NORPAX.

2. FLIP, various ship and aircraft measurements.

3. NORPAX Office, Scripps Institution of Oceanography, La Jolla, California, U.S.A.

Q. Project Prairie Grass

1. Project Prairie Grass was a field program designed to provide experimental data on the diffusion of a tracer gas over a range of 800 metres. Seventy experiments were conducted over the flat prairie in Nebraska, U.S.A. during July and August, 1956. The project was directed by the Atmospheric Analysis Laboratory of the Geophysics Research Directorate, Air Force Cambridge Research Centre.

2. Diffusion data, meteorological data.

3. M.L. Barad, Editor. Project Prairie Grass, A Field Program in Diffusion, Vols. I and II, 1958. Geophys. Res. Pap., 59, AFCRC, Bedford, Mass., U.S.A.

R. TROPEX-72

1. TROPEX-72 was carried out by the Soviet Union in the summer of 1972 in the tropical Atlantic Ocean. The experiment was carried out as a preparatory measure for participation in GATE. The tasks of the experiment were to test and evaluate observing systems; to develop methods for direct flux measurements; and to investigate the atmospheric motions including air-sea interactions on the small scale.

2. Standard meteorological data; radar observations; tetroons, tethered balloon and direct flux measurements in boundary layer; radiation measurements.

3. Petrossiants, M.A., 1973: Main objectives and some results of the National Expedition of the U.S.S.R. TROPEX-72. GATE, Rept. No. 2, World Meteor. Organ., Part I, 27 pp. Also other papers in same volume.

S. Wangara Experiment

1. The Wangara Experiment was conducted in July and August of 1967 at Hay, N.S.W., Australia. The objectives of the experiment were to find out whether processes on the large, meso-, or small synoptic scale contribute significantly to the vertical flux of angular momentum in the west wind belts of the world and to acquire data of use for more adequate description of the boundary layer.

2. Hourly pilot balloons, radiosondes, surface fluxes and profiles.

3. Clarke, R.H., 1970: Observational studies in the atmospheric boundary layer. Quart. J. Roy. Meteorol. Soc., 96, 91-114.

4. The Wangara Experiment: Boundary Layer Data, Editor - R.H. Clarke, Division of Meteorological Physics Technical Paper No. 19, C.S.I.R.O., Aspendale, Vic. 3195, Australia.

References

Axford, D.N., 1968: On the accuracy of wind measurements using an inertial platform in an aircraft, and an example of a measurement of the vertical mesostructure of the atmosphere. *J. Appl. Meteor.*, 7, 645-666.

Bean, B.R. and R.O. Gilmer, 1969: Comparison of barium fluoride humidity element with the microwave refractometer for studies of rapid fluctuations of atmospheric humidity. *Radio Science*, 12, 1155-1158.

Bean, B.R., R. Gilmer, R.L. Grossman, R. McGavin and C. Travis, 1972: An analysis of airborne measurements of vertical water vapor flux during BOMEX. *J. Atmos. Sci.*, 29, 860-869.

Bean, B.R., R.E. McGavin and B.D. Warner, 1973: A note on the FM-CW radar as a remote probe of the Pacific Trade-Wind inversion. *Boundary-Layer Meteor.*, 4, 201-210.

Bean, B.R., C.B. Emmanuel, R.O. Gilmer and R.E. McGavin, 1975: The spatial and temporal variations of the turbulent fluxes of heat, momentum and water vapour over Lake Ontario. *J. Phys. Oceanogr.*, 5, 532-540.

Benedetti-Michelangeli, G., F. Congeduti and G. Fiocco, 1972: Measurements of aerosol motion and wind velocity in the lower troposphere by Doppler optical radar. *J. Atmos. Sci.*, 29, 906-910.

Beran, D.W., 1974-75: Wind measurement with acoustic Doppler. *Atmos. Tech.*, 6, 94-99.

Brown, E.H. and F.J. Hall, 1978: Advances in atmospheric acoustics. *Rev. Geophys. and Space Phys.*, 16, 47-110.

Boston, N.E.J. and R.W. Burling, 1972: An investigation of high-wavenumber temperature and velocity spectra in air. *J. Fluid Mechanics*, 55, 473-492.

Browning, K.A., J.R. Starr and A.J. Whyman, 1973a: The structure of an inversion above a convective boundary layer as observed using high-power pulsed Doppler radar. *Boundary-Layer Meteor.*, 4, 91-112.

Browning, K.A., D. Beran, M.J.S. Quigley and C.G. Little, 1973b: Capabilities of Radar, Sodar and Lidar for measuring the structure and motion of the stably stratified atmosphere. *Boundary-Layer Meteor.*, 5, 195-200.

Bunker, A.F., 1955: Turbulence and shearing stress measured over the North Atlantic by an airplane acceleration technique. *J. Meteor.*, 12, 445-455.

Bunker, A.F., 1960: Heat and water-vapour fluxes in air flowing southward over the western North Atlantic Ocean. *J. Meteor.*, 17, 52-63.

Buck, A.L., 1973: Development of an improved Lyman-alpha hygrometer. *Atmos. Technol.*, 2, 43-46.

Burns, S.G., 1974-75: Boundary-layer instrumentations system. *Atmos. Tech.*, 6, 123-128.

Busch, N.E. and L. Kristensen, 1976: Cup anemometer overspeeding. *J. Appl. Meteor.*, 15, 1328-1332.

Camp, D.W., R.E. Turner and L.P. Gilchrist, 1970: Response tests of cup, vane and propeller wind sensors. *J. Geophys. Res.*, 75, 5265-5270.

Caughey, S.J. and R. Rayment, 1974: High-frequency temperature fluctuations in the atmospheric boundary layer. *Boundary-Layer Meteor.*, 5, 489-503.

Caughey, S.J. and J.C. Kaimal, 1977: Vertical heat flux in the convective boundary layer. *Q.J. Roy. Meteor. Soc.*, 103, 811-815.

Chadwick, R.B., K.P. Moran, R.G. Strauch, G.E. Morrison and W.C. Campbell, 1976a: A new radar for measuring winds. *Bull. Amer. Meteor. Soc.*, 57, 1120-1125.

Chadwick, R.B., K.P. Moran, R.G. Strauch, G.E. Morrison and W.C. Campbell, 1976b: Microwave radar wind measurements in clear air. *Radio Science*, 11, 795-802.

Cooney, J., 1977: Temperature profiles from laser radar measurements of rotational Raman and elastic scattering. Fourth Joint Conf., Sens. Env. Pollutants. Amer. Chem. Soc., Washington.

Deloach, R., A.L. Morris and R.B. McBeth, 1976: A combined boundary-profile and automated data-reduction and analysis system. *Boundary-Layer Meteor.*, 10, 455-463.

Donelan, M. and M. Miyake, 1973: Spectra and fluxes in the boundary layer of the Trade-Wind Zone. *J. Atmos. Sci.*, 30, 444-464.

Donelan, M.A., K.N. Birch and D.C. Beesley, 1974: Generalized profiles of wind speed, temperature and humidity. Proc. 17th Conf. on Great Lakes Res., Int. Assoc. Great Lakes Res., 778-788.

Dyer, A.J., B.B. Hicks and V. Sitaraman, 1970: Minimizing the levelling error in Reynolds stress measurements by filtering. *J. Appl. Meteor.*, 9, 532-534.

Dyer, A.J., B.B. Hicks and V. Sitaraman, 1971: Reply (to Kaimal and Haugen). *J. Appl. Meteor.*, 10, 339.

Dyer, A.J., 1975: Measurement of turbulent fluxes by fluxation and NIFTI techniques. *Atmos. Technology, 7*, 24-29.

Elliott, J.A., 1975: The measurement of pressure fluctuations in the atmospheric boundary layer. *Atmos. Technology, 7*, 30-32.

Estoque, M.A., J. Gross and H.W. Lai, 1976: A lake breeze over southern Lake Ontario. *Mon. Wea. Rev., 104*, 306-396.

Fiocco, G., G. Benedetti-Michelangeli, K. Maischberger and E. Madonna, 1971: Measurement of temperature and aerosol to molecule ratio in the troposphere by optical radar. *Nature, 229*, 78-79.

Frankel, M.S., N.J.F. Chang and M.J. Sanders, 1977: A high-frequency radio acoustic sounder for remote measurement of atmospheric winds and temperature. *Bull. Amer. Meteor. Soc., 58*, 928-934.

Friedman, H., G. Conrad and J. McFadden, 1970: ESSA Research Flight Facility aircraft participation in BOMEX. *Bull. Amer. Meteor. Soc., 51*, 822-834.

Friehe, C.A., C.H. Gibson, F.H. Champagne and J.C. LaRue, 1975: Turbulence measurements in the marine boundary layer. *Atmos. Technology, 7*, 15-23.

Garratt, J.R., 1975: Limitations of the eddy-correlation technique for the determination of turbulent fluxes near the surface. *Boundary-Layer Meteor., 8*, 255.

Gill, G.C., L.E. Olsson, J. Sela and M. Suda, 1967: Accuracy of wind measurements on towers or stacks. *Bull. Amer. Meteor. Soc., 48*, 665-674.

Gossard, E.E., J.H. Richter and D.R. Jensen, 1973: Effect of wind shear on atmospheric wave instabilities revealed by FM-CW radar observations. *Boundary-Layer Meteor., 4*, 113-132.

Grossman, R.L. and B.R. Bean, 1973: An aircraft investigation of turbulence in the lower layers of a marine boundary layer. NOAA Tech. Rept. ERL 291-WMPO 4, 166 pp.

Hardesty, R.M., P.A. Mandics, D.W. Beran and R.G. Strauch, 1977: The Dulles airport acoustic-microwave radar wind and wind shear measuring system. *Bull. Amer. Meteor. Soc., 58*, 910-918.

Haugen, D.A., J.C. Kaimal, C.J. Readings and R. Rayment, 1975: A comparison of balloon-borne and tower-mounted instrumentation for probing the atmospheric boundary layer. *J. Appl. Meteor., 14*, 540-545.

Hess, G.D., 1979: A guide to experimental design and analysis. The Planetary Boundary Layer, A.K. Blackadar (ed.), (To be published).

Hicks, B.B., 1972: Propeller anemometers as sensors of atmospheric turbulence. *Boundary-Layer Meteor.*, 3, 214-228.

Hooke, W.C., 1978: The implications of remote-sensing developments for air-pollution meteorology. Preprints WMO Symp. on Boundary-Layer Phys. Appl. Spec. Prob. of Air Poll., Norrköping, Sweden, WMO-No. 510, 259-270.

Horst, T.W., 1973: Corrections for response errors in a three-component propeller anemometer. *J. Appl. Meteor.*, 12, 716-725.

Huffaker, R.M., 1970: Laser Doppler detection systems for gas velocity measurement. *Appl. Optics*, 9, 1026.

Huffaker, R.M., 1974-75: CO_2 laser Doppler systems for the measurement of atmospheric winds and turbulence. *Atmos. Tech.*, 6, 71-76.

Hyson, P., 1972: Cup anemometer response to fluctuating wind speeds. *J. Appl. Meteor.*, 11, 843-848.

Izumi, Y. and M. Barad, 1970: Wind speeds as measured by cup and sonic anemometers and influenced by tower structure. *J. Appl. Meteor.*, 9, 851-856.

Jensen, N.O. and D.H. Lenschow, 1970: An observational investigation of penetrative convection. *J. Atmos. Sci.*, 35, 1924-1933.

Kaganov, E.I. and A.M. Yaglom, 1976: Errors in wind speed measurement by rotation anemometers. *Boundary-Layer Meteor.*, 10, 15-34.

Kaimal, J.C., 1968: The effect of vertical line averaging on the spectra of temperature and heat flux. *Quart. J. Roy. Meteor. Soc.*, 94, 149-155.

Kaimal, J.C. and D.A. Haugen, 1969: Some errors in the measurement of Reynolds stress. *J. Appl. Meteor.*, 8, 460-462.

Kaimal, J.C. and D.A. Haugen, 1971: Comments on "Minimizing the levelling error in Reynolds stress measurement by filtering". *J. Appl. Meteor.*, 10, 337-339.

Kaimal, J.C., J.C. Wyngaard, Y. Izumi and O.R. Coté, 1972: Spectral characteristics of surface layer turbulence. *Quart. J. Roy. Meteor. Soc.*, 98, 563-589.

Kaimal, J.C., 1975: Sensors and techniques for direct measurements of turbulent fluxes and profiles in the atmospheric surface layer. *Atmos. Technology*, 7, 7-14.

Kaimal, J.C., J.C. Wyngaard, D.A. Haugen, O.R. Coté, Y. Izumi, S.J. Caughey and C.J. Readings, 1976: Turbulence structure in the convective boundary layer. *J. Atmos. Sci.*, 33, 2152-2169.

Kaimal, J.C. and D.A. Haugen, 1977: An acoustic Doppler sounder for measuring wind profiles in the lower boundary layer. *J. Appl. Meteor.*, 16, 1298-1305.

Kerman, B.R., 1976: A note on the interpretation of acoustic sounder returns. *Boundary-Layer Meteor.*, 10, 303-310.

Kjelaas, A.G. and G.R. Ochs, 1974: Study of divergence in the boundary layer using optical propagation techniques. *J. Appl. Meteor.*, 13, 242-248.

Kropfli, R.A. and N.M. Kohn, 1977: Persistent horizontal rolls in the urban mixing layer as revealed by dual Doppler radar. Preprints, Seventh Conf. on Inadvertent and Planned Weather Modification; Amer. Meteor. Soc., Boston, Mass.

Lawrence, R.S., G.R. Ochs and S.F. Clifford, 1972: Use of scintillations to measure average wind across a light beam. *Appl. Optics.*, 11, 239-243.

Lemon, L.R., R.J. Donaldson, D.W. Burgess and R.A. Brown, 1977: Doppler radar application to severe thunderstorm study and potential real-time warning. *Bull. Amer. Meteor. Soc.*, 58, 1187-1193.

Lemone, M.A. and W.T. Pennell, 1976: The relationship of Trade-Wind cumulus distribution to subcloud layer fluxes and structure. *Mon. Wea. Rev.*, *104*, 524-539.

Lenschow, D.H., 1971: Vanes for sensing incidence angles of the air from an aircraft. *J. Appl. Meteor.*, 10, 1339-1343.

Lenschow, D.H., 1973: Two examples of planetary boundary layer modification over the Great Lakes. *J. Atmos. Sci.*, 30, 568-581.

Lenschow, D.H. and W.T. Pennell, 1974: On measurements on in-cloud and wet-bulb temperatures from an aircraft. *Mon. Wea. Rev.*, 102, 447-454.

Lenschow, D.H., 1975: The use of aircraft for probing the atmospheric boundary layer. *Atmos. Technology*, 7, 47-49.

Lenschow, D.H., 1976: Estimating updraft velocity from an airplane response. *Mon. Wea. Rev.*, 104, 618-627.

Lenschow, D.H. and E.M. Agee, 1976: Preliminary results from AMTEX. *Bull. Amer. Meteor. Soc.*, 57, 1346-1355.

Lilly, D.K. and D.H. Lenschow, 1971: Aircraft measurements of the atmospheric mesoscales using an inertial reference system. Facilities for Atmospheric Research, 19, 2-8.

Little, G.C., 1972: Status of remote sensing of the troposphere. *Bull. Amer. Meteor. Soc.*, *53*, 936-949.

Lo, A.K., 1977: Boundary layer flow over gentle curvilinear topography with a sudden change in surface roughness. *Q.J. Roy. Meteor. Soc.*, *103*, 199-210.

Lumley, J.L. and H.A. Panofsky, 1964: The Structure of Atmospheric Turbulence. Wiley: Interscience, London, 239 pp.

MacCready, P.B., 1966: Mean wind speed measurements in turbulence. *J. Appl. Meteor.*, *5*, 219-225.

MacPherson, J.I., 1973: A description of the NAE T-33 turbulence research aircraft, instrumentation and data analysis. Nat. Res. Council of Canada, DME/NAE Quart. Bull., 1972(4), 31 pp.

MacPherson, J.I., 1974: Results of intercomparison flights between the NAE T-33 and the NCAR Buffalo atmospheric research aircraft. Nat. Res. Council of Canada, Aeronautical Rept. LR-577, 30 pp.

McAllister, L.G., J.R. Pollard, A.R. Mahoney and P.J.R. Shaw, 1969: Acoustic sounding - a new approach to the study of atmopsheric structure. Proc. IEEE, 57, 579-587.

McBean, G.A., 1972: Instrument requirements for eddy correlation measurements. *J. Appl. Meteor.*, *11*, 1078-1084.

McBean, G.A., 1975: Comments on 'limitations of the eddy-correlation technique for the determination of turbulent fluxes near the surface.' *Boundary-Layer Meteor.*, *9*, 361-362.

McBean, G.A. and R.D. Paterson, 1975: Variations of the turbulent fluxes of momentum, heat and moisture over Lake Ontario. *J. Phys. Oceanogr.*, *5*, 523-531.

McBean, G.A. and J.I. MacPherson, 1976: Turbulence above Lake Ontario: velocity and scalar statistics. *Boundary-Layer Meteor.*, *10*, 181-198.

Merceret, F.J., 1976a: The turbulent microstructure of Hurricane Caroline (1975). *Mon. Wea. Rev.*, *104*, 1297-1307.

Merceret, F.J., 1976b: Airborne hot-film measurements of the small-scale structure of atmospheric turbulence during GATE. *J. Atmos. Sci.*, *33*, 1739-1746.

Mickle, R.E. and D.S. Davison, 1979: Differences in kilometer-scale wind and humidity boundary layer structure at GATE. *ATMOSPHERE-OCEAN*, 17, (to be published).

Miyake, M., M. Donelan and Y. Mitsuta, 1970: Airborne measurement of turbulent fluxes. *J. Geophys. Res.*, 75, 4506-4518.

Miyake, M. and G.A. McBean, 1970: The measurement of vertical humidity transport over land. *Boundary-Layer Meteor.*, 1, 88-101.

Morris, A.L. and F. Hall, 1974-75: Remote sensing of the atmosphere by sound. *Atmos. Tech.*, 6, 84-93.

Morris, A.L., D.B. Call and R.B. McBeth, 1975: A small tethered balloon sounding system. *Bull. Amer. Meteor. Soc.*, 56, 964-969.

Noonkester, V.R., D.R. Jensen, J.H. Richter, W. Viezce and R.T.H. Collis, 1974: Concurrent FM-CW radar and lidar observations of the boundary layer. *J. Appl. Meteor.*, 13, 249-256.

Panofsky, H.A., 1965: Re-analysis of Swinbank's Kerang observations: flux of heat and momentum in the planetary boundary layer. Rept., Dept. of Meteor., Penn. St. Univ., 66-76.

Pasquill, F., 1972: Some aspects of boundary layer description - Presidential Address. *Q.J. Roy. Meteor. Soc.*, 98, 469-494.

Paulson, C.A., 1970: The mathematical representation of wind speed and temperature profiles in the unstable atmospheric surface layer. *J. Appl. Meteor.*, 8, 857-861.

Paulson, C.A., E. Leavitt and R.G. Fleagle, 1972: Air-sea transfer of momentum, heat and water determined from profile measurements during BOMEX. *J. Phys. Ocean.*, 2, 487-497.

Pennell, W.T. and M.A. Lemone, 1974: An experimental study of turbulence structure in the fair-weather Trade-Wind boundary layer. *J. Atmos. Sci.*, 31, 1308-1323.

Polavarapu, R.J. and R.E. Munn, 1967: Direct measurement of vapour pressure fluctuations and gradients. *J. Appl. Meteor.*, 6, 669-706.

Polavarapu, R.J., 1972: Direct measurement of vapour pressure and its fluctuations using fine thermocouples. *Quart. J. Roy. Meteor. Soc.*, 98, 198-205.

Rao, K.S., J.C. Wyngaard and O.R. Coté, 1974: The structure of the two-dimensional internal boundary layer over a sudden change of surface roughness. *J. Atmos. Sci.*, 31, 738-746.

Rayment, R., 1973: An observational study of the vertical profile of the high-frequency fluctuations of wind in the atmospheric boundary-layer. *Boundary-Layer Meteor.*, 3, 284-300.

Rayment, R. and C.J. Readings, 1974: A case study of the structure and energetics of an inversion. *Quart. J. Roy. Meteor. Soc.*, 100, 221-233.

Readings, C.J. and H.E. Butler, 1972: The measurement of atmospheric turbulence from a captive balloon. *Meteor. Mag.*, 3, 284-300.

Readings, C.J., E. Golton and K.A. Browning, 1973: Fine-scale structure and mixing within an inversion. *Boundary-Layer Meteor.*, 4, 275-288.

Readings, C.J., 1975: The use of a tethered kite-balloon to probe the lowest levels of the atmosphere. *Atmos. Technology*, 7, 33-37.

Schwerdtferger, P., 1976: Physical Principle of Micrometeorological Measurements. Elsevier, Amsterdam, 113 pp.

Schwiesow, R.L. and R.E. Cupp, 1976: Remote Doppler velocity measurements of atmospheric dust devil vortices. *Appl. Optics.*, 15, 1-2.

Sheih, C.M., H. Tennekes and J.L. Lumley, 1971: Airborne measurements of the small-scale structure of atmospheric turbulence. *Phys. Fluids*, 14, 201-215.

Strauch, R.G., V.E. Derr and R.E. Cupp, 1971: Atmospheric temperature measurement using Raman backscatter. *Appl. Optics.*, 10, 2665-2669.

Strauch, R.G., V.E. Derr and R.E. Cupp, 1972: Atmospheric water vapour measurement by Raman lidar. *Remote Sens. Env.*, 2, 101-108.

Taylor, P.A., 1969: The planetary boundary layer above a change in surface roughness. *J. Atmos. Sci.*, 26, 432-440.

Taylor, P.A., 1970: A model of airflow above changes in surface heat flux, temperature and roughness for neutral and unstable conditions. *Boundary-Layer Meteor.*, 1, 18-39.

Taylor, P.A., 1971: Airflow above changes in surface heat flux, temperature and roughness; an extension to include the stable case. *Boundary-Layer Meteor.*, 1, 474-497.

Thompson, N., 1972: Turbulence measurements over the sea by a tethered-balloon technique. *Quart. J. Roy. Meteor. Soc.*, 98, 745-762.

Vugts, H.F. and J.A. Businger, 1977: Air modification due to a step change in surface temperature. *Boundary-Layer Meteor.*, *11*, 295-306.

Warner, J., 1971: Observations of the eddy fluxes of heat and vapor over the sea. *Q.J. Roy. Meteor. Soc.*, *97*, 540-547.

Warner, J., 1972: The structure and intensity of turbulence in air over the sea. *Q.J. Roy. Meteor. Soc.*, *98*, 175-186.

Weill, A., W. Brandes, C. Oberto and P. Bares, 1976: Anémomètre a fil chaud embarquable sur ballon captif. *J. Physique*, *C1*, 173-174.

Wesley, M.L. and B.B. Hicks, 1975: Comments on 'limitations of an eddy correlation technique for the determination of the carbon dioxide and sensible heat fluxes'. *Boundary-Layer Meteor.*, *9*, 363-367.

Wesley, M.L., 1976: A comparison of two optical methods for measuring line averages of thermal exchanges above warm water surfaces. *J. Appl. Meteor.*, *15*, 1177-1188.

Wieringa, J., 1967: Evaluation and design of wind vanes. *J. Appl. Meteor.*, *6*, 1114-1122.

Wieringa, J., 1972: Tilt errors and precipitation effects in trivane measurements of turbulent fluxes over open water. *Boundary-Layer Meteor.*, *2*, 406-426.

Wyngaard, J.C., Y. Izumi and S.A. Collins, 1971: Behavior of the refractive-index structure parameter near the ground. *J. Opt. Soc. Amer.*, *61*, 1646-1650.

Wyngaard, J.C., 1973: On surface-layer turbulence. Workshop on Micrometeorology, D.A. Haugen, (ed.), Boston, Amer. Meteor. Soc., 101-149.

Wyngaard, J.C., J.T. Bauman and R.A. Lynch, 1974: Cup anemometer dynamics. Flow: Its Measurement and Control in Science and Industry, Vol. 1, (ed. by R.B. Dowdell), Instrument Soc. of Amer., 701-708.

Wyngaard, J.C., J.C. Kaimal, G.R. Ochs, R.J. Hill and D.C. Sorenson, 1978: An optical heat flux experiment. Proc. Fourth Symp. Meteor. Obs. and Instrum., Amer. Meteor. Soc., Boston.

Wyngaard, J.C., W.T. Pennell, D.H. Lenschow and M.A. Lemone, 1978: The temperature-humidity covariance budget in the convective boundary layer. *J. Atmos. Sci.*, *35*, 47-58.

Wucknitz, J., 1977: Disturbance of wind profile measurements by a slip mast. *Boundary-Layer Meteor.*, 11, 115-170.

Yap, D., T.A. Black and T.R. Oke, 1974: Calibration and tests of a yaw sphere-thermometer system for sensible heat flux measurements. *J. Appl. Meteor.*, 13, 40-45.

Yelagina, L.G., 1969: The relation of the frequency spectra of humidity fluctuations to moisture fluxes. *Izv. Atmos. Oceanic Phys.*, 5, 621-622.

Yokoyama, O., 1969: Measurements of wind fluctuations by a vane mounted on a captive balloon cable. *J. Meteor. Soc. Japan*, 47, 159-166.

Yokoyama, O., M. Gamo and S. Yamamoto, 1973: Aircraft measurement on the structure of planetary boundary layer. Scientific Report, Second AMTEX Study Conf., AMTEX Report No. 3, 56-68.

CHAPTER 6 - APPLICATIONS

6.1 Estimation of Surface Fluxes

6.1.1 Introduction

A good estimate of the rate of the evaporative loss of water from a reservoir would enable more optimum utilization of the available water for irrigation, power generation, human consumption and other uses. Similarily, a good estimate of the rate of sensible heat loss from a high latitude lake would enable a better prediction of the time of freeze-up. For these and many other practical applications it is desirable to have good techniques for estimating the transfers of energy, mass and momentum between the atmosphere and the underlying surface.

Measurement systems to more directly determine the fluxes have been described in Chapter 5. Because evaporation is, in a practical sense, the most important of these fluxes it has received considerable attention, particularly from the agricultural and hydrological points of view. Chapter 3 of WMO Technical Note 97 (1968) was devoted to the estimation of evaporation for efficient use of water for irrigation. Two more recent Technical Notes (No. 119, the Application of Micrometeorology to Agricultural Problems, 1972; and No. 126, Comparison between Pan and Lake Evaporation, 1973) and a monograph of the American Meteorological Society (Vol. 6, No. 28, Agricultural Meteorology, (1965) have considered the same problem. In addition there exist several textbooks, primarily on agricultural meteorology, that provide information on the practical estimates of surface fluxes (e.g. Rose, 1966; Smith, 1975). And, finally, there is the excellent recent review by Webb (1975) on the evaporation from catchments. Other reviews and references are included in the bibliography. In view of the available information on this subject, this section shall be short and quite selective. No attempt will be made to review the variety of empirical methods that utilize either climatological or very limited data. The review will be restricted to techniques that have a sound physical basis and utilize recent advances in boundary-layer meteorology.

6.1.2 Profile Techniques

The profile techniques rely on the existence of known relationships between the surface fluxes and the profiles of the relevant mean parameter in the layer of air immediately above the surface. As yet the techniques are applicable only over extensive, uniform, horizontal surfaces. In order to establish the profile at least three, and preferably more, levels of measurements are required. These levels of measurement must be high enough that the flux is transported by fully turbulent processes and at the same time be low enough that the flux is not yet changing sufficiently with height. Measurements between a few tens of centimetres and ten to twenty metres will usually satisfy these requirements.

The basic assumption of the profile technique is that the flux, F_A, is related to the gradient of the profile by:

$$F_A = - \rho K_A \, \partial \bar{A}/\partial z$$

It is further assumed that the eddy diffusivity, K_A, can be written in the form:

$$K_A = k u_* z/\Phi_A(z/L)$$

following from Monin-Obukhov similarity theory (see Chapter 1).

For fluxes of momentum, $\tau = \rho u_*$, sensible heat, $H = - \rho C_p \, u_* \, \theta_*$ and moisture, $E = - \rho \, u_* \, q_*$ there are three important parameters u_*, θ_*, q_* and these can be related to the gradients by:

$$\frac{kz}{u_*}\frac{\partial U}{\partial z} = \Phi_M\left(\frac{Z}{L}\right) \qquad U = \text{mean horizontal wind}$$

$$\frac{kz}{\theta_*}\frac{\partial \theta}{\partial z} = \Phi_H\left(\frac{Z}{L}\right) \qquad \theta = \text{mean potential temperature} \qquad (6.1.1)$$

$$\frac{kz}{q_*}\frac{\partial Q}{\partial z} = \Phi_E\left(\frac{Z}{L}\right) \qquad Q = \text{mean specific humidity}$$

It follows that:

$$U(z_2) - U(z_1) = U_2 - U_1 = \frac{u_*}{k}\left[\ln\frac{z_2}{z_1} - \psi_M\left(\frac{Z}{L}\right)\right] \qquad (6.1.2)$$

where

$$\psi\left(\frac{Z}{L}\right) = \int_{z_1/L}^{z_2/L} \frac{1-\Phi\,(\eta)}{\eta}\,d\eta \qquad \text{and } z_1, z_2 \text{ are any two levels} \qquad (6.1.3)$$

Similar equations can be written for $\theta(z)$ and $Q(z)$. The equations for u_*, θ_* and q_* and the resulting fluxes are:

$$u_* = \frac{k\,(U_2-U_1)}{\ln\frac{z_2}{z_1} + \psi_M\left(\frac{Z}{L}\right)}$$

(6.1.4)

$$\theta_* = \frac{k\,(\theta_2 - \theta_1)}{\ln \frac{z_2}{z_1} + \psi_H\left(\frac{z}{L}\right)}$$

$$q_* = \frac{k\,(Q_2 - Q_1)}{\ln \frac{z_2}{z_1} + \psi_E\left(\frac{z}{L}\right)}$$

$$\tau/\rho = u_*^2 \qquad H/\rho C_p = -\,u_*\,\theta_* \qquad E/\rho = -\,u_*\,q_*$$

These equations give the formal relationships between the fluxes and profiles. Note that to evaluate the heat flux one must first evaluate the momentum flux by know $U(z)$ and $\psi_M(z/L)$ but in order to know z/L both the heat flux and the momentum flux are needed. The method by which the equations can be solved depends on the functional form of $\psi_M(z/L)$. Since z/L depends on the surface fluxes which are to be determined, it is preferable to use the Richardson Number, Ri, which depends on the profiles, as a stability parameter. The Ri is directly related to z/L by:

$$Ri = \frac{g}{T} \frac{\frac{\partial \Theta}{\partial z}}{\left(\frac{\partial U}{\partial z}\right)^2} = \frac{z}{L} \frac{\Phi_H}{\Phi_M^2} \tag{6.1.5}$$

In principle, if the functions Φ_H and Φ_M are known then z/L can be computed from Ri.

Paulson (1970) has outlined the steps to compute the fluxes from profile information on wind and temperature. In brief, the steps are:

1. Compute the Ri using either a finite difference approximation to gradients, i.e.

$$\left.\frac{\partial \bar{A}}{\partial z}\right|_{z'} \approx \frac{\bar{A}_2 - \bar{A}_1}{(z_1 z_2)^{1/2}} \left[\ln\left(z_2/z_1\right)\right]^{-1} \qquad z' = (z_1 z_2)^{1/2} \tag{6.1.6}$$

   or by fitting higher order polynomials to the data and differentiating which is preferable.

2. From Ri compute z/L, ψ_M and ψ_H.

3. Regress U and θ versus $\ln z + \psi_M$ and $\ln z + \psi_H$ respectively. The slopes will be u_*/k and θ_*/k respectively. The intercepts will be $(-u_*/k) \ln z_o$ and $(-\theta_*/k) \ln z_T + \theta\,(z_T)$.

An alternate approach, as proposed by Van Ulden (unpublished) can be used if the roughness length z_o is known. It has the advantages that the heights of wind and temperature measurement need not be the same and only one wind speed and two temperature measurements is needed. The flux equations can then be written as:

$$u_* = \frac{k\,U_1}{\ln \frac{z_1}{z_o} + \psi_M\left(\frac{z_1}{L}\right)}$$

$$\theta_* = \frac{k\,[\theta_3 - \theta_2]}{\ln \frac{z_3}{z_2} + \psi_H\left(\frac{z_3}{L}\right) - \psi_H\left(\frac{z_2}{L}\right)} \qquad (6.1.7)$$

The heights z_1, z_2 and z_3 can be anywhere in the 0.5 m to 10 m range. The roughness length, z_o, can be obtained by means of extensive windspeed profile measurements during periods of high windspeeds when neutral stability is closely approximated or by the method suggested by Wieringa (1976) which uses gust-factor measurements to deduce z_o at non-ideal site.

The procedure to compute the fluxes is to compute u_* and θ_* from (6.1.7) using a first guess of L; then recomputing u_* and θ_* using the L based on the first guess u_* and θ_*. This interation is repeated until u_* and θ_* are computed to the required accuracy (this usually is accomplished in about 5 iterations for better than 0.5% accuracy).

Regardless of which of the above methods is used, the accuracy of the computation will depend upon:

1. the accuracy of the data;

2. the applicability of the theory to the data, and

3. the knowledge of ψ_M, ψ_H etc.

Observational methods have been presented in Chapter 5, including a discussion of errors. The applicability of the theory depends primarily on the site and the conditions under which the data were collected. It is not possible to provide general answers to this question.

Our knowledge of the functional forms and constants in the surface-layer flux-gradient relations has been much improved over the last two decades but there are still some outstanding problems. For many purposes the moisture flux or evaporation rate is the most important. Unfortunately it also is the least well known of the momentum, sensible heat and moisture flux group. If the humidity content of the air is small then the buoyancy forces are almost entirely due to temperature differences. This is common at mid-latitudes and over dry land at all latitudes. In this case the moisture is carried by the turbulence and does not significantly effect the turbulence; i.e., it is passive. To evaluate the evaporation rate it is necessary to measure the temperature and wind profiles in order to evaluate the Richardson Number and the specific humidity profile to get the evaporation rate. This is assuming that Φ_E is known, a point to be returned to.

Over warm water bodies the moisture content of the air is frequently large enough that the humidity differences between parcels of air significantly effect their relative buoyancy. In this case the moisture plays an active role in the turbulent motions. The Richardson Number must then be redefined as being in terms of the gradient of virtual potential temperature instead of potential temperature. The functional form for Φ_E may then also be different.

The question arises now as to how well are the Φ's known. There have been several studies to determine Φ_M and Φ_H and a few for Φ_E and, unfortunately, the results are not conclusive. Some recent reviews of the subject are by Dyer (1975), Yaglom (1977) and McBean (1976).

For unstable stratification the following forms for Φ_M have been proposed:

(a) $$\Phi_M = \left(1 - \sigma z/_L\right)^{-1/4} \tag{6.1.8}$$

Dyer and Hicks (1970), (hereafter DH) and Paulson (1970) both proposed $\sigma = 16$ while Businger, Wyngaard, Izumi and Bradley (1971, hereafter BWIB) found $\sigma = 15$. Both were based on good experimental data. The apparent agreement is, however, tempered by the fact that DH used $k = 0.41$, Paulson used $K = 0.40$ and BWIB used $k = 0.35$.

(b) $$\Phi_M\,(Ri) = \left(1 - 16\ Ri\right)^{-1/3} \tag{6.1.9}$$

This result, due to Pruitt, Morgan and Laurence (1973, hereafter PBL), differs considerably from (a) but is also based on good experimental data. A value of $k = 0.42$ was found.

For stable stratifications there are again two formulations, that arise from the same data sets as the unstable formulations.

(a) $$\Phi_M = 1 + \sigma z/_L \tag{6.1.10}$$

Webb (1970) proposed $\sigma = 5.2$ ($k = 0.4$) but the data set did not include direct measurements of u_*. BWIB found $\sigma = 4.7$ ($k = 0.35$). Numerous other studies have reported values of $\sigma \sim 5$.

(b) $$\Phi_M = \left(1 + b\ Ri\right)^{1/n} \tag{6.1.11}$$

Different values have been proposed for b and n. For example:

	n	$^1/_n$	b	
(i)	2	1/2	10	Oke (1970a), Rossby and Montgomery (1935)
(ii)	3	1/3	16	PML
(iii)	4	1/4	18	Panofsky et al., (1960). This is a variation KEYPS formula.
			14 - 24	Sellers (1965).

PML compared these three formulations and found that for $0 < Ri \lesssim 0.1$ that (i) and (ii) agree well with (iii) with $b < 20$. For $Ri > 0.1$ (i) increases much faster than (ii) and $b \sim 38$ in (iii) is needed to have agreement with (ii).

The formulations for Φ_H are no better determined than for Φ_M. For unstable stratifications the most acceptable form is:

$$\Phi_H = A_H \left(1 - \sigma z/_L\right)^{-1/2} \qquad (6.1.12)$$

DH found $\sigma = 16$, $A_H = 1$ while BWIB found $\sigma = 9$, $A_H = 0.74$. Since $K_H/K_M = \Phi_M/\Phi_H$ the BWIB result is $K_H/K_M = 1.35$ for neutral stability while the DH result gives 1.0. Although a value of 1.0 has many supporters and follows from the Reynolds analogy, a value of greater than 1.0 had been found in many studies previous to BWIB (e.g., see Lumley and Panofsky, 1964, pp. 105-106). Hogstrom's (1974) results for undisturbed flow conditions at his "typical" agricultural site were also in good agreement with BWIB for Φ_M, Φ_H and k in unstable and neutral conditions. McBean and Miyake (1972) inferred that $K_H/K_M > 1$ for near neutral conditions by examining the correlation coefficient functions.

For stable stratifications a form similar to that for Φ_M has been proposed by several researchers; i.e.,

$$\Phi_H = A_H + \alpha z/_L \qquad z/_L > 0 \qquad (6.1.13)$$

Webb (1970), proposed $A_H = 1$ and $\alpha = 4.2$, (the same as for Φ_M) while BWIB were again somewhat different; $A_H = 0.74$, $\alpha = 4.7$ (α the same as for Φ_M). Although Hogstrom's results agreed quite well with BWIB for unstable conditions, his results for stable conditions did not agree nor even follow similarily. Thus, the necessity for an "ideal" site may be more important for stable conditions than for unstable conditions. Stable conditions over land usually occur at night when the wind speeds are also light. Oke (1970b) investigated the temperature profiles on calm clear nights and found that the temperature minimum was frequently raised above the surface and that radiative flux divergences were important.

Because the DH and BWIB data were analysed using different values of k their agreement on the formula for Φ_M and to a lesser extent for Φ_H is somewhat fortuitous. In Figure 6.1.1, the results of DH, BWIB, PBL and Webb are plotted in a way independent of k; i.e., k/Φ vs $z/_{L_*}$ where $L_* = kL$. The intercept at $z/L_* = 0$

is the value of k used. There are obviously large differences in the formulations. However, what is important from the practical point of view is the error in the deduced fluxes. Given a set of wind and temperature observations at several levels what are the differences between the momentum and heat fluxes determined from the BWIB formulation relative to, say, the DH formulation. Lo and McBean (1978) have investigated this question and found that, for u_*^2, differences of 18% between DH and BWIB for typical unstable conditions and 20% between PML and BWIB for neutral conditions, rising to nearly 50% for z/L = -.75. For stable conditions the PML case deviated greatly from the others. These differences can be compared with a difference of about 22% caused by a +5% deviation in the profile for the top levels and a -5% deviation for the bottom levels (using the BWIB formulation).

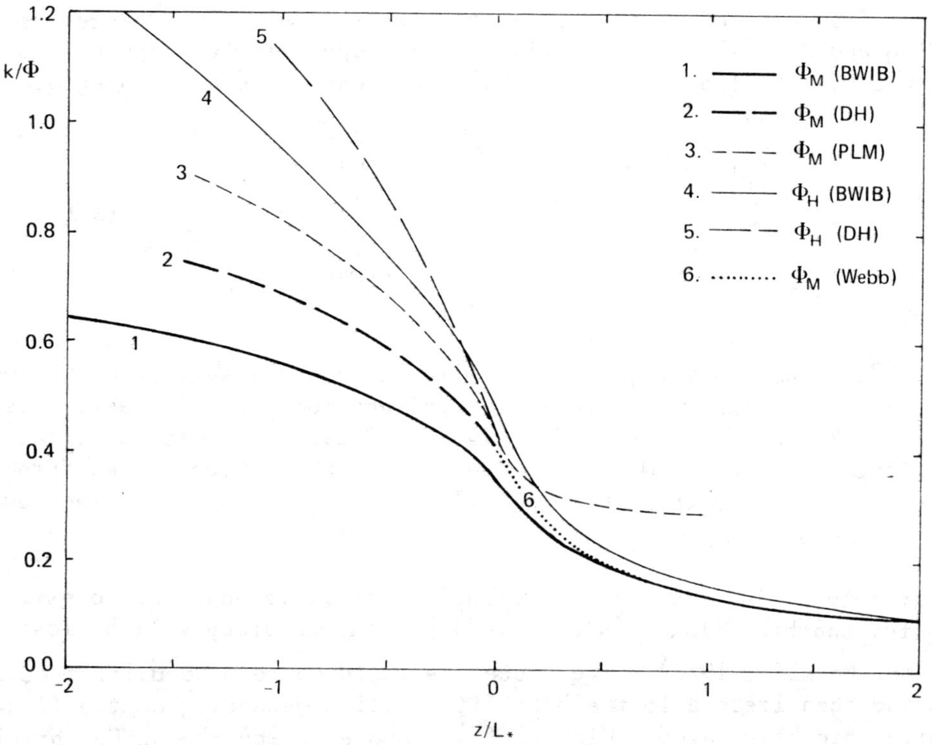

Figure 6.1.1 Comparison of non-dimensional wind shear and temperature gradients. $z/L_* = z/kL$. 1 - Φ_M (BWIB); 2 - Φ_M (DH); 3 - Φ_M (PML); 4 - Φ_H (BWIB); 5 - Φ_H (DH); 6 - Φ_M (Webb). For the PML curve it was assumed that Ri = z/L for unstable conditions and Ri = z/L (1 + 5.2 z/L) for stable conditions.

The flux-gradient relation for moisture, Φ_E, is not well established but it is generally accepted as being $\Phi_E = \Phi_H$ (e.g. DH). However, in some recent work, this relationship has been questioned. Hogstrom (1974) found that $\Phi_E \sim \Phi_H$ for strong instability but $\Phi_E \sim \Phi_M = 1.35\ \Phi_H$ for neutrality. The Φ_E also displayed a larger scatter than the data for Φ_H or Φ_M. PML did not measure the heat flux directly so Φ_H results were not published but their Φ_E values agree approximately with the Φ_H's of BWIB and DH for moderate instabilities but are between them for near neutral stabilities. For stable conditions their Φ_E does not agree with the BWIB Φ_H. They found that $\Phi_E/\Phi_M = .88$ at neutral stability. These results show that the dependence of the water vapour flux on stability is still not clear. However, it is reasonable to expect Φ_E to be near but likely larger than Φ_H.

Warhaft (1976) has shown that the ratio Φ_E/Φ_H $(= K_H/K_E)$ will be significantly different from unity when the correlation between temperature and humidity fluctuations is small and a) when either θ or q (but not both) is acting as a passive contaminant, or b) when $\partial\theta/\partial z$ and $\partial Q/\partial z$ have opposite signs.

For free-convection stability conditions an alternate procedure, as proposed by Ting and Hay (1975), is available. The approximate range of applicability is $0.1 < -z/L < 2.0$. The free convective profile equations can be integrated to give

$$\theta = 3C_1 \left(H/\rho C_p\right)^{2/3} \left(g/T_o\right)^{-1/3} z^{-1/3} + \text{constant}$$

(6.1.14)

$$U = -3C_2\, u_*^2 \left(Hgz/\rho C_p\, T_o\right)^{-1/3} + \text{constant}$$

Ting and Hay (1975) and Sundararajan and Macklin (1976) have shown that the matching of these equations to profiles give reliable heat and momentum flux estimates. Based on the AFCRL Kansas Experiment data (Izumi, 1971) the values for C_1 and C_2 are 1.0 and 1.65 respectively. Note that the heat flux can be computed directly from the temperature profile while both profiles are needed to obtain the momentum flux.

If only a wind profile is available then it is possible to estimate both the momentum and heat fluxes (Klug, 1967) but the accuracy will be less (Sundararajan and Macklin, 1976). The procedure would be to make a first estimate of the fluxes and then iterate to the best fit solution depending on the flux-gradient relationship being used. Klug (1967) proposed using the KEYPS formulation but it would be possible to use any other set of formulae.

6.1.3 Bulk Transfer Formulae

In certain circumstances the profile techniques of the previous section can be greatly simplified and yet still give useful flux estimates. Consider the equation for the kinematic momentum flux with $z_1 = z_o$ and $z_2 = 10$ m:

$$u_*^2 = \frac{k^2\, U_{10}^{\ 2}}{\left[\ln \frac{10}{z_o} + \psi_M \left(\frac{10}{L}\right)\right]} = C_D\, U_{10}^{\ 2} \qquad (6.1.15)$$

with $C_D = k^2 / \left[\ln \frac{10}{z_o} + \psi_M\ \frac{10}{L}\right]^2$ = the drag coefficient.

The heat flux can be written as:

$$H/\rho C_p = -u_* \theta_* = \frac{k^2 U_{10} (\Theta_o - \Theta_{10})}{[\ln \frac{10}{z_o} + \psi_M (\frac{10}{L})] [\ln \frac{10}{z_T} + \psi_H (\frac{10}{L})]}$$

$$= C_H U_{10} (\Theta_0 - \Theta_{10}) \qquad (6.1.16)$$

and similarity for moisture flux

$$E/\rho = -u_* q_* = C_E U_{10} (Q_o - Q_{10}) \qquad (6.1.17)$$

The coefficients C_D, C_H, and C_E include the effects of surface roughness, atmospheric stability and any possible wind effect other than indicated. This kind of approach has been quite successful over water surfaces but not over land or vegetated surfaces. For a water surface the surface temperature and specific humidity (assuming a saturated value at the surface temperature) can be measured. For land or vegetated surfaces this is much more difficult. Also the variations of surface roughness and stability are much greater for land or vegetated surfaces.

The drag coefficient, C_D, has been most extensively studied and the general consensus is that it is about 1.5×10^{-3} with a possible error of $\pm$ 50%. It is probable that C_D varies with wind speed and its exact variation is important because of the relative importance of high wind periods to the overall momentum budget. Smith and Banke (1975) suggest $10^3 C_D = 0.63 + 0.66 U_{10} \pm 0.23$ and Wieringa (1974) found $10^3 C_D = 0.86 + 0.058 U_{10}$. Others (e.g., Hicks, 1972; DeLeonibus, 1971) have also found C_D increasing with U_{10} but Dunkel et al., (1974) and others report no significant change. For very light winds it is possible that C_D decreases with U_{10} increases in accordance with the aerodynamically smooth theories (Hicks et al., 1974). The stability dependence of C_D has been investigated by DeLeonibus (1971) but needs further study. Since the stability over water is usually near neutral the stability effect is not too important.

The dependence of C_D on surface roughness and wave parameters is more important. Kitaigorodski (1970) has shown that the stress for constant wind speed depends on the waves through the ratio of c/u_* where c is the phase speed of the waves. Generally the stress decreases as c/u_* increases. This concept is coupled with theories of wave generation. If the waves are moving faster than the wind then little wave generation is taking place. It is important, therefore, that studies of C_D include measurements of wave characteristics. Hicks et al. (1974) have also shown that C_D is less for shallow water than for deep water; an effect that is related to the waves. The drag coefficient, however, is a good first approximation for use in evaluating the local drag.

Most of the studies referred to above involved eddy correlation measurements of local momentum flux to the surface. The surface drag can also be estimated from wind set up. Using this technique, Wieringa (1974) found $C_D \sim 0.0024$ in agreement with some earlier results while Donelan et al. (1975) found $C_D \sim 0.0015$.

There are considerable difficulties in deducing these values but there are indications that effective drag coefficient over extended water masses may be larger than the point measurements indicate. Numerical modellers of ocean or lake circulations have frequently had to use large (2 to 3 x 10^{-3}) values of C_D to get the water movements to agree with observations. This may be related to the fact that most eddy correlation measurements are made in near steady-state seas and hence may not be representative of many situations.

For the transfer of heat and water vapour, values of C_H and C_E similar to that for C_D are usually advocated (Businger, 1975). In most studies the number of observations and the range of stabilities and wind speeds has not been large so that only average values are meaningful. Pond et al. (1974) found $C_H = C_E = 0.0015$; Hasse (197C), $C_H = 0.001$; Coantic (1974), $C_H = C_E = 0.0013$ for $U_{10} < 10$ ms $^{-1}$ and $0.001 + 0.00005\, U_{10}$ for $10 < U_{10} < 20$ ms^{-1}; Smith (1974), $C_H = C_E = 0.0012$; and Hicks (1972), $C_H = .0014$. Frieche and Schmidt (1975) used the results of Smith and Banke (1975), Dunckel et al. (1974) and their own to obtain:

(i) for $0 < U_{10}\Delta T < 25$

$$\overline{w\theta} = 0.002 + (0.97 \times 10^{-3})\ U\Delta T \text{ in } (\text{ms}^{-1})\ (^{\circ}\text{k})$$

(ii) $U\Delta T > 25$

$$\overline{w\theta} = (1.46 \times 10^{-3})\ U\Delta T$$

(iii) $U\Delta T < 0$

$$\overline{w\theta} = 0.0026 + (0.86 \times 10^{-3})\ U\Delta T$$

(iv) $\overline{wq} = (1.32 \times 10^{-3})\ U\Delta q$ in $(\text{ms}^{-1})\ (\text{gm}^{-3})$ for all data (no high wind speed cases).

Measurements over the tropical ocean (Pond et al., 1971 ; Dunckel et al., 1974) have led to very high apparent values for C_H. Usually $U\Delta T$ is very small and the heat flux may be determined by other factors, such as its relationship to evaporation. Although the sensible heat flux loss from the ocean may be small compared to the latent heat flux loss, it can still be very important in determining the surface layer stability. As yet the problem of heat flux transfers over the tropical ocean is not entirely resolved. For example, the study of Friehe and Schmidt (1975) indicated that the observed "cold spikes" (e.g., Pond et al., 1971) may be instrumental effects and these would lead to anomously high values for C_H. It is hoped that several recent field experiments such as GATE, AMTEX, JASIN (see Chapter 5) will resolve some of these problems.

6.1.4 Planetary Boundary Layer Techniques

The bulk transfer techniques discussed in the previous section rely on knowledge of the winds, temperature and humidity at the surface and at 10 m height. In this section we consider relationships between the fluxes and the differences in wind, temperature and humidity over the whole planetary

$$C_f = \frac{u_*}{G} = k\ \{a^2 \left(\frac{h}{L}\right) + \left[\ln\left(\frac{h}{z_o}\right) - b\left(\frac{h}{L}\right)\right]^2\}^{-1/2}$$

$$\sin \alpha = - \frac{a(h/L)}{k} \frac{u_*}{G} \text{ sign } f$$

$$C_H = \frac{-H}{k\rho C_p u_* (\theta_h - \theta_o)} = \alpha_H \left[\ln \left(\frac{h}{z_o}\right) - c \left(\frac{h}{L}\right) \right]^{-1} \quad (6.1.18)$$

$$C_E = \frac{-E}{k\rho u_* (q_h - q_o)} = \alpha_D \left[\ln \left(\frac{h}{z_o}\right) - d \left(\frac{h}{L}\right) \right]^{-1}$$

At present the experimental evaluations of a, b, c and d and their variations with stability are not well known (see Chapter 2). So the only approximate evluation of the surface fluxes can be made. Deardorff (1972) proposes an alternate method which combines both surface and planetary boundary layer formulations. The method was proposed for use in general circulation models but could also be used for direct applications of computing the surface fluxes.

Again, as the data from more recent experiments is analysed, it is expected that these techniques will be improved so more reliable flux estimates are possible.

6.1.5 Turbulence Measurements Techniques

Studies of the constant-flux surface layer have shown that there exist useful relationships between the statistics of the turbulent fluctuations and the surface fluxes. One approach, based on the turbulent energy budget and Kolmogoroff's hypotheses, has been called variously, the ϕ_{11} technique (Miyake et al., 1970), the spectral density technique (Hicks and Dyer, 1972) and the inertial-dissipation technique (Champagne et al., 1977). In this technique the turbulent energy budget equation is used to derive a relationship between the energy dissipation rate ε and the momentum flux u_*^2. The complexity of the relationship has varied with different authors, depending on their view of the importance of the individual terms in the energy budget. The dissipation is then related to the spectral energy density of the longitudinal (usually) component in the inertial subrange. The resulting equation can be written in the form

$$u_*^2 = C f^{2/3} f\phi_{11}(f) F(z/L) \quad (6.1.19)$$

where

C is a numerical constant depending on the von Karman constant and the Kolmogoroff constant

f is the non-dimensional frequency

$f\phi_{11}(f)$ is the spectral energy density of the longitudinal component

F(z/L) is a function of stability depending on the form of the turbulent energy budget used.

The advantage of this technique is that only the fluctuations of one component for a specified frequency band in the inertial subrange need be measured. This is particularly important for measurement from ships, aircraft and other moving platforms because the frequency band can usually be chosen so as to avoid platform motion effects.

Similar equations can be written relating θ_* and q_* to the spectral densities of temperature and humidity fluctuations. As yet these approaches have not been widely used mainly because of the uncertainties in the inertial subrange constants.

Tillman (1972) has proposed a method based on the skewness and standard deviations of temperature fluctuations. It is applicable to dry unstable conditions. The surface stress can also be determined from the maximum of the vertical-velocity spectrum, as shown by Moravek et al., (1976). Williams (1977) developed a modification of the dissipation technique that requires measurements only of the mean and fluctuating temperatures at two levels within the constant flux layer. Both momentum and heat fluxes can be deduced. Each of these techniques relies on the Monin-Obukhov hypothesis and are only applicable for steady-state conditions over homogeneous terrain.

References

Businger, J.A., J.C. Wyngaard, Y. Izumi and E.F. Bradley, 1971: Flux-profile relationships in the atmospheric surface layer. *J. Atmos. Sci.*, 28, 181-189.

Businger, J.A., 1975: Interactions of sea and atmosphere. *Rev. of Geophys. and Space Phys.*, 13, 720-725.

Champagne, F.H., C.A. Friche, J.C. LaRue and J.C. Wyngaard, 1977: Flux measurements, flux estimation techniques, and fine-scale turbulence measurements in the unstable surface layer over land. *J. Atmos. Sci.*, 34, 515-530.

Coantic, M., 1974: Formules empiriques d,Evaporation. Note de la Convention. CNEXO/IMST No. 74/951.

Deardorff, J.W., 1972: Numerical investigation of neutral and unstable planetary boundary layers. *J. Atmos. Sci.*, 29, 91-115.

De Leonibus, P.S., 1971: Momentum flux and wave spectra observations from an ocean tower. *J. Geophys. Res.*, 76, 6506-6527.

Dunckel, M., L. Hasse, L. Krugermeyer, D. Schriever and J. Wucknitz, 1974: Turbulent fluxes of momentum, heat and water vapor in the atmospheric surface layer at sea during ATEX. *Boundary-Layer Meteorol.*, 6, 81-106.

Dyer, A.J. and B.B. Hicks, 1970: Flux-gradient relationships in the constant flux layer. *Quart. J. Roy. Meteor. Soc.*, 96, 715-721.

Dyer, A.J., 1974: A review of flux-profile relationships. *Boundary-Layer Meteor.*, 7, 363-372.

Frenzen, Paul, 1973: The observed relation between the Kolmogorov and von Karman constants in the surface boundary layer. *Boundary-Layer Meteor.*, 3, 348-358.

Friehe, C.A. and K.F. Schmidt, 1975: Parameterizations of air-sea interface fluxes of sensible heat and moisture by the bulk aerodynamic formulas. To be published.

Hasse, L., 1970: On the determination of the vertical transports of momentum and heat in the atmospheric boundary layer at sea. Tech. Rept. No. 188, School of Oceanography, Oregon State Univ., 55 pp.

Hicks, B.B., 1972: Some evaluations of drag and bulk transfer coefficients over water bodies of different sizes. *Boundary-Layer Meteorol.*, 3, 194-213.

Hicks, B.B., R.L. Drinkrow and G. Grouse, 1974: Drag and bulk transfer coefficients associated with a shallow water surface. *Boundary-Layer Meteor.*, 6, 287-298.

Hicks, B.B. and A.J. Dyer, 1972: The spectral density technique for the determination of eddy fluxes. *Q. J. Roy. Meteor. Soc.*, 98, 838-844.

Hogstrom, U., 1974: A field study of the turbulent fluxes of heat water vapour and momentum at a "typical" agricultural site. *Quart. J. Roy. Meteorol. Soc.*, 100, 624-639.

Izumi, Y., 1971: Kansas 1968 field program data report. Env. Res. Pap., 379, AFCRL-72-0041, Air Force Cambridge Research Laboratories, Bedford Massachusetts.

Kitaigorodskii, S.A., 1970: The physics of air-sea interaction. Leningrad, 284 pp.

Klug, W., 1967: Determination of turbulent fluxes of heat and momentum from the wind profile. *Q. J. Roy. Meteor. Soc.*, 93, 101-104.

Lumley, J.L. and H.A. Panofsky, 1964: The structure of atmospheric turbulence. Interscience, New York 239 pp.

Lo, A.K. and G.A. McBean, 1978: On the relative errors in methods of flux calculations. *J. Appl. Meteor.*, 17, (Nov. issue).

McBean, G.A. and M. Miyake, 1972: Turbulent transfer mechanisms in the atmospheric surface layer. *Quart. J. Roy. Meteor. Soc.*, 98, 392-398.

Miyake, M., M. Donelan, G. McBean, C. Paulson, F. Badgley and E. Leavitt, 1970: Comparison of turbulent fluxes over water determined by people and eddy correlation techniques. *Q. J. Roy. Meteor. Soc.*, 96, 132-137.

Moravek, D., H.A. Panofsky and A. Weber, 1976: Determination of surface stress from vertical-velocity spectra. *Q. J. Roy. Meteor. Soc.*, 102, 260-263.

Oke, T.R., 1970a: Turbulent transport near the ground in stable conditions. *J. Appl. Meteor.*, 9, 778-786.

Oke, T.R., 1970b: The temperature profile near the ground on calm clear nights. *Quart. J. Roy. Meteor. Soc.*, 96, 14-23.

Panofsky, H.A., A.K. Blackadar and G.E. McVehil, 1960: The diabatic wind profile. *Quart. J. Roy. Meteor. Soc.*, 86, 390-398.

Paulson, C.A., 1970: The mathematical representation of wind and temperature profiles in the unstable surface layer. *J. Appl. Meteor.*, 9, 857-861.

Pond, S., G.T. Phelps, J.E. Paquin, G.A. McBean and R.W. Stewart, 1971: Measurements of the turbulent fluxes of momentum moisture and sensible heat over the ocean. *J. Atmos. Sci.*, 28, 901-

Pond, S., D.B. Fissel and C.A. Paulson, 1974: A note on bulk aerodynamic coefficients for sensible heat and moisture fluxes. *Boundary-Layer Meteor.*, 6, 333-340.

Pruitt, W.O., D.L. Morgan and F.J. Lourence, 1973: Momentum and mass transfers in the surface boundary layer. *Quart. J. Roy. Meteor. Soc.*, 99, 370-386.

Rose, C.W., 1966: Agricultural Physics. Pergamon, Oxford, 226 pp.

Rossby, C.G. and R.B. Montgomery, 1935: The layer of frictional influence in wind and ocean current. Papers Phys. Ocean Met., MIT Woods Hole Ocean. Institut., 3:3.

Sellers, W.D., 1965: Physical Climatology. University of Chicago Press, Chicago and London.

Smith, L.P., 1975: Methods in Agricultural Meteorology. Elsevier, Amsterdam, 210 pp.

Smith, S.D., 1974: Eddy flux measurements over Lake Ontario. *Boundary-Layer Meteor.*, 6, 235-256.

Smith, S.D. and E.G. Banke, 1975: Variation of the sea surface drag coefficient with wind speed. *Quart. J. Roy. Meteor. Soc.*, 101, 665-670.

Sundararajin, A., and S.A. Macklin, 1976: Comments "On the heat flux and friction velocity in free convection near the ground." *J. Atmos. Sci.*, 33, 715-718.

Tillman, J.E., 1972: The indirect determination of stability, heat and momentum fluxes in the atmospheric boundary layer from simple scaler variables during dry unstable conditions. *J. Appl. Meteor.*, 11, 783-792.

Ting, C.L. and D.R. Hay, 1975: On the heat flux and friction velocity in free convection near the ground. *J. Atmos. Sci.*, 32, 637-638.

Warhaft, Z., 1976: Heat and moisture flux in the stratified boundary layer. *Q. J. Roy. Meteor. Soc.*, *102*, 703-707.

Webb, E.K., 1970: Profile relationships: the log-linear range and extension to strong stability. *Quart. J. Roy. Meteor. Soc.*, *96*, 67-90.

Webb, E.K., 1975: Evaporation from catchments. Prediction in Catchment Hydrology, Aust. Acad. of Sci., 203-236.

Wieringa, J., 1974: Comparison of three methods for determining strong wind stress over Lake Flevo. *Boundary-Layer Meteorol.*, *7*, 3-19.

Wieringa, J., 1976: An objective exposure correction method for average wind speeds measured at a sheltered location. *Q. J. Roy. Meteor. Soc.*, *102*, 241-253.

Williams, R.M., 1977: Estimation of momentum and heat fluxes in the constant flux layer from temperature measurements. *Boundary-Layer Meteor.*, *12*, 299-305.

Yaglom, A.M., 1977: Comments on wind and temperature flux-profile relationships. *Boundary-Layer Meteor.*, *11*, 89-102.

6.2 Aeronautical Meteorology

6.2.1 Introduction

Meteorology of the boundary layer is an important ingredient in the design and operation of aircraft and aerospace vehicles. So much so that most meteorological services owe their initial existence to supporting the aviation industry. Advances in understanding and utilizing knowledge resulting from research in atmospheric science coupled with advances in aeronautical technology have combined to significantly reduce the hazards of air travel. At the same time the growing level of engineering sophistication has presented a new set of problems demanding both additional and refined knowledge of the atmospheric boundary layer.

The purpose of this Section is to acquaint the interested meteorologist with an overview of the interactions of aeronautics and atmospheric science in dealing with problems to aircraft and aerospace vehicles in the atmospheric boundary layer. The orientation of the discussion will follow the line of the earlier chapters of this Technical Note where the status and thrust of recent boundary layer research has been reviewed. The discussion is not meant to be exhaustive but rather to catch the highlights of the interaction between the two fields and to stress the importance of new findings in the PBL. Hopefully the bibliography will lead the interested meteorologist to an in-depth examination of particular problems. In particular reference is made to WMO Technical Note 95 on Aeronautical Meteorology. The major difficulties of a meteorological nature to aeronautical operations in the planetary boundary layer evolve from the structure and turbulence of the wind and from restrictions to visibility. Winds and turbulence cause aircraft buffetting, altering its lift/drag characteristics and its stability, to say nothing of the discomfort to passengers. The aeronautical engineering community is particularly concerned with cross-winds on approach, departure or surface manoeuvers and with near-surface vertical shears. Consequently the research results on the surface layer are particularly important. However, knowledge of the characteristics of the Ekman or outer layer are also important. In addition, the transient and localized nature of particular mesoscale structures such as thunderstorms or topographically induced flows are also of concern and require alternative formulations.

The existence of fog or low cloud are the main restrictors to visibility. These problems require a concentrated effort of research on such fields as the effects of radiation, cloud microphysics and thermodynamics of clouds and fog in addition to boundary-layer structure. In recent years several undesirable effects of aircraft operations, such as noise and air pollution, have also required meteorological expertise. The attenuation of sound in air by a combination of temperature, humidity and turbulence effects should be a factor in planning air terminal operations. Similarily, the possibility of high concentrations of exhaust products is dictated by boundary-layer considerations.

Rather than review all these topics, which would require considerable development outside the scope of this Technical Note, it has been decided to restrict the discussion to the directly applicable results of recent advances in our understanding of the PBL and to wind effects on aircraft and aerospace vehicles.

6.2.2 Surface Winds and Turbulence

6.2.2.1 Some Current Problem Areas

A recent survey by Houbolt (1973) of designers and operators of aircraft provides an indication of some of the problems facing the aeronautical community. Included among these are: controlled hovering of helicopters in moderate to heavy, gusty winds; stability and control of short take-off and landing (STOL) aircraft, particularly in confined areas; stability effects related to wind shear; pilot-induced instabilities; control of flexible wing aircraft in turbulence; flow around obstacles; and the creation of wind-tip vortices. Further discussion of these problems may be found in the proceedings or reports on conferences on aerospace and aeronautical meteorology (e.g. American Meteorological Society, 1970, 1972, 1974, 1976; Royal Aeronautical Establishment, 1961; WMO Technical Note 95) and in the extensive bibliography of Houbolt (1973).

Another area of considerable concern to aeronautical engineers has been the effects of buffetting of aerospace vehicles, with the resulting structural fatigue and stability and navigation problems. Design philosophy and techniques are discussed in Ryan and King (1969) and Daniels (1971). The approach taken by the aeronautical community has necessarily simplified the specification of atmospheric inputs because of the incomplete nature of our understanding of the PBL and its interaction with aircraft. However, recent advances in PBL understanding, as reported in earlier chapters of this Note, now allow for a more systematic incorporation of wind and turbulence effects (see, for example, Luers, 1973; Kerman, 1977a). Further, the basic meteorological parameters controlling the turbulent structure are now more easily derivable from standard surface and upper-air observations (Blackadar et al., 1974).

6.2.2.2 Formulation of Aeronautical Response to Atmospheric Winds and Turbulence

In many applications, the aeronautical response or output, y(t), to an atmospheric input, u(t), can be expressed in the form of a generalized differential equation:

$$L\,\{y(t)\} = u(t) \tag{6.2.1}$$

where L is a differential operator, or in an equivalent integral equation form:

$$y(t) = \int_0^{\infty} K(t,\tau)\; u(\tau)\,d\tau \tag{6.2.2}$$

where $K(t,\tau)$ is the appropriate function relating input and response. By response, one implies such behaviour as the displacement, velocity, or acceleration of a component or totality of the aeronautical structure, and by input, usually the wind velocity or accelerations. The functions y(t) and u(t) are stochastic variables, linked by a deterministic transformation denoted by L and K.

These latter are considered to be known to the aeronautical engineer (they will vary with aircraft type). The meteorological input is, of course, the atmospheric characteristics represented by u(t). As with most stochastic processes, the analysis is considerably simpler if the stochastic input processes is assumed to be stationary and homogeneous, that is, its probablistic structure or statistics do not change with time or space. Then the probablistic structure of the output or aircraft response is expressible in terms of the probablistic structure of the atmospheric input. For example, the interrelationship of the moments, defined in general as:

$$E\{\xi^n\} = \int_{-\infty}^{\infty} \xi^n(t)\ dt \qquad (6.2.3)$$

becomes

$$E\{y^n\} = \int_{-\infty}^{\infty} \cdots \int_{-\infty}^{\infty} \{K(\tau_1) \ldots K(\tau_n) E\ u^n(t-\tau_1-\tau_2 \ldots -\tau_n)\} d\tau_1 \ldots d\tau_n \qquad (6.2.4)$$

Usually only the first (mean, n = 1) and second (variance, n= 2) order responses of the output are required. The second order relationship is more clearly expressed in spectral form as:

$$\phi_y(\omega) = |\hat{K}(\omega)|^2\ \phi_u(\omega) \qquad (6.2.5)$$

that is, the output spectrum, $\phi_y(\omega)$, is a filtered (by $|\hat{K}(\omega)|^2$) output of the input atmospheric spectrum, $\phi_u(\omega)$. Extensions of (6.2.5) for non-linear responses, requiring additional moments of the input probablistic structures, have been built around the Wiener-Hermite functional representation (Kerman, 1977a) of which the preceding is a subset.

To illustrate the application of (6.2.5), consider the lift, or equivalently, the vertical acceleration, averaged over a wing surface, of an aircraft in gusty winds. The mean lift is a function of the mean wind speed component in the aircraft's longitudinal axis while the spectrum of lift fluctuations consists of weighting the spectral energy available in the atmospheric turbulence by the lift response mode or modes of the aircraft. The analogy is similar to a multi-mode electronic filter passing energy or power from the input side to the output side.

Unfortunately, although the above approach is widely applicable, it is insufficient to cover several important areas of concern. In the case of aircraft with high-wing loadings, such as STOL aircraft, and of aerospace vehicles extending through a significant portion of the boundary layer, the joint behaviour of turbulence acting simultaneously on various parts of the structure becomes important (Case, 1968; Daniels, 1971). The significant variables to such an analysis are the coherence and phase difference between velocity components with various vector separations (Dutton et al., 1971). Observations to-date on the spatial structure are not as extensive as other boundary-layer measurements nor have such measurements that do exist been as completely analysed, principally because of the data processing costs and the difficulties involved. However, some empirical information on coherence and phase does exist (e.g., Davenport, 1961; Dutton et al., 1971; Ropeleski et al., 1973; Pielke and Panofsky, 1970; Berman and Stearns, 1977) and do, for example, demonstrate the effects of atmospheric stability. Extended versions of (6.2.5), to account for joint occurrences of turbulent winds, can then produce a similar relationship between input and output spectra. The concept of exceedence is also useful in dynamic load problems. For example, how often does the turbulent velocity in a certain direction exceed a threshold value. Some data on this are available (e.g., McBean, 1974) and the problem has been analysed theoretically by Dutton (1970).

In addition to the dynamic response, the static loadings of an aircraft structure are also of interest in applications. For example, the engineer is concerned with the probability of the wind exceeding a certain value, at a given height, with a given confidence and risk level. Such information at the design stage and for operations will allow for quantitative decision making. The theories of repeated trials and extremes (Gumbel, 1958) can be applied to relate the value of the wind, the corresponding time periods and the risk. These statistics will vary with climatological and topographical factors pertaining to each location.

The above basic tools of interaction between meteorolgy and aeronautics indicate the need for various statistics of the turbulent wind field. A review of some of the relevant aspects of boundary layers is presented next.

The above basic tools of interaction between meteorology and aeronautics indicate the need for various statistics of the turbulent wind field. A review of some of the relevant aspects of boundary layers is presented next.

6.2.2.3 Structure of the Surface Boundary Layer

Surveys of the structure of the atmospheric boundary layer (for example, Lumley and Panofsky, 1964; Haugen (ed.), 1973; Monin and Yaglom, 1971) indicate the importance of the Monin-Obukhov similarity theory for the atmospheric surface layer (see Chapter 1). Recent observational studies (e.g., Businger et al., 1971; McBean et al., 1971; Kaimal et al., 1976) have improved our understanding of the validity and applicability of the theory.

In the surface boundary layer the characteristic velocity, u_*, temperature, T_*, and length, z, (see Chapter 1) have been used to establish non-dimensional functions for mean shear, temperature gradient, variance and other turbulent statistics as functions of the stability parameter, z/L. For example, for mean shear:

$$\frac{kz}{u_*}\frac{\partial U}{\partial z} = \phi_M(z/L) = (1-15z/L)^{-1/4} \quad -1 < z/L \leq 0 \qquad (6.2.6)$$

$$= 1 + 4.7\, z/L \quad 0 \leq z/L < 0.5$$

(Businger et al., 1971) and for the longitudinal (along-wind) component as approximately:

$$\frac{\overline{u^2}}{\sigma_u^2} \doteq 6.25\ (1-0.4z/L)^2 \qquad (6.2.7)$$

(Kerman, 1977b). A discussion of these useful empirical functions can be found in the above references (see also, Yaglom, 1977, regarding the wind shear and temperature gradient relations).

The M-O theory has been extended to the frequency, ω, or equivalently the wavenumber, k, domain by the normalization by z/U or by z, as appropriate, to form:

$$f = \omega z/U = kz/2\pi \qquad (6.2.8)$$

For example, the spectrum of vertical velocity variance, w, can be represented as (Kaimal et al., 1972):

$$f\phi_w(f,\ z/L) = 1.15(f/f_m)/\left[1 + 2(f/f_m)^{5/3}\right]$$

where f_m = 0.56 for neutral conditions.

Note that this relationship is consistent with the Kolmogorov (1941) hypotheses in that the slope of ϕ_w is $f^{-5/3}$ for f >> 1. Other empirical relationships in the past have frequently neglected this consistency in favour of mathematical convenience. Additional empirical formulae for spectra and cross spectra can be found in Kaimal et al., (1972), Wyngaard and Coté (1972) and others referred to in Haugen (1973). An application utilizing these spectra is described in Kerman (1977b).

The benefit of utilizing the above methodology and empirical formulations is that the number of basic unknowns for the engineer is reduced to two (u_* and T_*) plus some characteristic boundary conditions. For example, if u_* and T_* and the surface roughness are known then the velocity profile can be constructed (note that surface roughness is not needed for the shear profile). From u_* and the stability (which depends on u_*, T_* and z) the variances of the turbulent velocity components and their spectra (using the wind speed to convert to frequency) can be deduced. Although our knowledge of surface layer relationships is not yet perfect (see Chapter 6.1) the resulting estimates provide useful quantitative values for the aeronautical engineering applications.

The use of M-O similarity for the surface boundary layer, unfortunately, as yet cannot be applied to a variety of problems because of spatial inhomogeneities or nonstationarities. Transient events such as rainstorms with their associated local accelerations may exclude the existence of a constant flux layer or may cause systematic deviations from the "ideal case" parameterizations discussed above. The flow around obstacles, such as buildings, may, under certain conditions, be treated by standard engineering techniques (for example, Frost, 1973; Frost et al., 1977) as a distortion of the basic meteorological flow and as an introduction of turbulence variance in a selected wavelength regime depending on the shedding characteristics of the obstacle. Flow over simple changes in surface roughness (e.g., Taylor 1969; Rao et al., 1974) or over idealized topographical configuration (Taylor, 1977a,b) have been the object of considerable modelling research in the recent past. From these type of models it is possible to estimate u_* and T_* during the adjustment process to a downstream condition. However, as yet, it is not known whether the preceding empirical relations will still be valid nor to what extent. Regarding the problems associated with nonstationarity very little is as yet known.

In summary it should be noted that surface layer structure for conditions of horizontal homogeneity and time stationarity is sufficiently well known to permit the construction of useful wind-turbulence models in support of aeronautics. Problems still remain with the full specification of the turbulence spatial correlations, with the accurate assessment and incorporation of the non-linear, non-Gaussian aspects of turbulence, and with the effects of inhomogeneity and nonstationarity.

6.2.3 Structure of the Ekman Layer

Our understanding of the Ekman layer structure and turbulence characteristics, although significantly advanced in recent years, is still somewhat fragmentary (Chapters 1 and 2), particularly for stable stratifications. Nevertheless, considerable information is available for the construction of engineering models. The matching of Ekman and surface layer properties leading to the resistance laws (see Chapters 1 and 2) allows for the computation of u_* and T_* if the surface roughness, geostrophic velocity and temperature deficit across the total boundary layer are known. Alternatively they can be computed from knowledge of the inversion height, surface wind and radiation data. As outlined previously this would then allow the determinations of wind structure and statistics near the surface. The application could be either for single events for which no special boundary layer data are available, such as an aircraft accident (Kerman, 1974) or on a climatological basis, such as when the probablistic distribution of variance or shear is needed. To-date no authors seemed to have utilized the latter concept.

As was indicated in Chapter 1, we have reasonably good knowledge of the structure of the unstable, convective boundary layer. The relevant basic parameters, the surface heat flux and the inversion height, are combined to give a characteristic velocity, w_*, and the similarity structure proceeds in much the same manner as for the surface layer. The similarity functions can be deduced from the experimental data (e.g., Kaimal et al., 1976; Panofsky et al., 1977). From these data it is possible to construct a model for the convective boundary layer that is almost complete as that for the surface layer.

However, in several regards, our knowledge is incomplete. For example, clarification is needed of the turbulent and gravity wave structure near the capping conversion (Metcalfe, 1974) which may seriously affect aircraft. Also the presence of roll structures and coalescing thermal bubbles (Lemone, 1973; Hall et al., 1975) need further study in order to be incorporated into models of use for applications to aeronautics.

The stable boundary layer has not proved as amenable to similarity solutions as the convective boundary layer (see Chapters 1 and 2) because of its weak vertical interaction. The consequence of its inability to efficiently transport vertical turbulent momentum leads to marked wind shears in the vertical, which can be of considerable aeronautical importance. The shearing may occur in the boundary layer itself or at the top in an inversion layer where it may be associated with significant turbulence due to breaking gravity waves (Richter, 1969). This has been discussed in Chapter 1.

Although the relationships for the PBL are not well established there have been some extensive experimental programmes to characterize turbulence for aeronautical modelling (e.g., Gault, 1967; the Lo-Locat studies in the USA). The parameterization arising from the Lo-Locat programme have been set into broad characterizations of surface type and stability and are sufficient for overall averages and estimates of variability but are insufficient for constructing models as detailed as those for the surface and convective boundary layers.

The neutral PBL is of interest when it is associated with extensive mixing and turbulence associated with high winds.

Panofsky (1973) has reviewed the structure of what he calls the "tower layer", about the lowest 150 m. He shows that the logarithmic wind profile appears to be valid for heights well above the surface layer. This may be due to the counter-balancing effects of the change in stress with height and the Coriolis influence. In Chapter 2 a more complete review of neutral Ekman layers is given.

6.2.4 Summary

Wind and turbulence play an important role in aeronautical operations, both affecting their efficiency and safety. The principles and formulations of boundary-layer meteorology can provide valuable assistance in bettering their operations. As yet boundary-layer research cannot provide all the answers but, present knowledge, as outlined in this section and given in more detail in other parts of this Technical Note, can provide in many instances the required information and guidance.

References

Berman, S. and C.R. Stearns, 1977: Near-earth turbulence and coherence measurements at Aberdeen Proving Ground, Md. *Boundary-Layer Meteor.*, 11, 485-506.

Blackadar, A.K., H.A. Panofsky and F. Fiedler, 1974: Investigation of the turbulent wind field below 500 feet altitude at the Eastern Test Range, Florida, U.S. National Aeronautics and Space Administration, NASA CR-2438.

Businger, J.A., J.C. Wyngaard, Y. Izumi and E.F. Bradley, 1971: Flux-profile relationships in the atmospheric surface layer. *J. Atmos. Sci.*, 28, 181-189.

Case, E.R., 1968: Development of a low altitude gust model and its application to STOL aircraft response studies. De Havilland Aircraft of Canada, DHC-DIR-68-15.

Daniels, G.E., editor, 1971: Terrestrial environment (climatic) criteria guidelines for use in space vehicle development. U.S. National Aeronautics and Space Administration, NASA TM X-64589.

Davenport, A.G., 1961: The spectrum of horizontal gustiness near the ground in high winds. *Quart. J. Roy. Met. Soc.*, 87, 194-211.

Dutton, J.A., 1970: Effects of turbulence on aeronautical systems. Progress in Aerospace Sciences, 11, 67-109.

Dutton, J.A., H.A. Panofsky, D.C. Deaven, B.R. Kerman and V. Mirabella, 1971: Statistical properties of turbulence at the Kennedy Space Center for Aerospace Vehicle Design. NASA CR-1889.

Frost, W., 1973: Review of data and prediction techniques for wind profiles around man-made surface obstructions. AGARD Cinference Proceedings No. 140 on Flight in Turbulence, Advisory Group for Aerospace Research and Development, 7, Rue Ancelli, 92200, Neuilly Sur Seine. France.

Frost, W., G. Fichtl, J.R. Connell and M.L. Hutto, 1977: Mean horizontal wind profiles measured in the atmospheric boundary layer about a simulated block building. *Boundary-Layer Meteor.*, 11, 135-146.

Gault, J.D., 1967: Low altitude atmospheric turbulence analysis methods. *Canadian Aero Space J.*, 13.

Gumbel, E.J., 1958: *Statistics of Extremes*. Columbia Univ. Press, 385 pp.

Hall, F.F., J.G. Edinger and W.D. Neff, 1975: Convective plumes in the planetary boundary layer investigated with an acoustic echo sounder. *J. Applied Met.*, 14, 513-523.

Haugen, D.A., editor, 1973: *Workshop on Micrometeorology*. Am. Met. Soc., 392 pp.

Houbolt, J.C., 1973: Survey on effect of surface winds on aircraft design and operation and recommendations for needed wind research. NASA CR-7360.

Kaimal, J.C., J.C. Wyngaard, Y. Izumi and O.R. Coté, Y. Izumi, J.S. Caughey and C.J. Readings, 1976: Turbulence structure in the convective boundary layer. *J. Atmos. Sci.*, 33, 2152-2169.

Kerman, B.R., 1974: An application of boundary layer similarity theories to an aircraft accident investigation. *Boundary-Layer Met.*, 7, 53-63.

Kerman, B.R., 1977a: The discrete turbulent gust concept. *Boundary-Layer Meteor.*, in press.

Kerman, B.R., 1977b: Multi-component linear digital simulation of atmospheric boundary layer turbulence. *J. Industrial Aerodynamics*, in press.

Lemone, M.A., 1973: The structure and dynamics of horizontal roll vortices in the planetary boundary layer. *J. Atmos. Sci.*, 30, 1077-1091.

Luers, J.K., 1973: A model of wind shear and turbulence in the surface boundary layer. NASA CR-2288.

Lumley, J.L. and H.A. Panofsky, 1964: *Structure of Atmospheric Turbulence*. Interscience-Wiley, 239 pp.

McBean, G.A., R.W. Stewart and M. Miyake, 1971: The turbulent energy budget near the surface. *J. Geophy. Res.*, 76, 6540-6549.

McBean, G.A., 1974: The turbulent transfer mechanisms: a time domain analysis. *Q. J. Roy. Meteor. Soc.*, 100, 53-66.

Metcalf, J.I., 1974: Gravity waves in a low-level inversion. *J. Atmos. Sci.*, 32, 351-361.

Monin, A.S. and A.M. Yaglom, 1971: Statistical Fluid Mechanics. MIT Press, 769 pp.

Monin, A.S. and S.S. Zilintinkevich, 1974: Similarity theory and resistance laws for the planetary boundary layer. *Boundary-Layer Meteor.*, 7, 391-397.

Panofsky, H.A., 1973: Tower Meteorology. *Workshop on Micrometeorology*, (D.A. Haugen, ed.,), Amer. Met. Soc., 151-176.

Panofsky, H.A., H. Tennekes, D.H. Lenschow and J.C. Wyngaard, 1977: The characteristics of turbulent velocity components in the surface layer under convective conditions. *Boundary-Layer Meteor.*, 11, 355-361.

Pielke, R.A. and H.A. Panofsky, 1970: Turbulence characteristics along several towers. *Boundary-Layer Meteor.*, 1, 115-130.

Rao, K.S., J.C. Wyngaard and O.R. Cote, 1974: The structure of the two-dimensional internal boundary layer over a sudden change of surface roughness. *J. Atmos. Sci.*, 31, 738-746.

Richter, J.H., 1969: High resolution tropospheric radar sounding. *Radio Sci.*, 4, 1261-1268.

Ropeleski, C.F., H. Tennekes and H.A. Panofsky, 1973: Horizontal coherence of wind fluctuations. *Boundary Layer Meteor.*, 5, 353-363.

Ryan, R.S. and A.W. King, 1969: The influential aspects of atmospheric disturbances on space vehicle design using statistical approaches for analyses. NASA TN-D-4969.

Taylor, P.A., 1970: A model of air flow above changes in surface heat flux, temperature and roughness for neutral and unstable conditions. *Boundary-Layer Meteor.*, 1, 18-39.

Taylor, P.A., 1977a: Some numerical studies of surface boundary layer flow above gentle topography. *Boundary-Layer Meteor.*, 11, 439-465.

Taylor, P.A., 1977b: Numerical studies of neutrally stratified planetary boundary-layer flow above gentle topography, I: two-dimensional cases. *Boundary-Layer Meteor.*, 12, 37-60.

Wyngaard, J.C. and O.R. Coté, 1972: Co-spectral similarity in the atmospheric surface layer. *Quart. J. Roy. Met. Soc.*, 98, 590-603.

Yaglom, A.M., 1977: Comments on wind and temperature flux-profile relationships. *Boundary-Layer Meteor.*, 11, 89-102.

6.3 Urban Boundary Layer

6.3.1 Introduction

Recent years have seen a noticeable growth in the number of publications, conferences and symposia on urban climatology (Chandler, 1968, 1970, 1976; Oke, 1974). Research on atmospheric characteristics in urbanized areas is presently underway, the essence of which concerns designing and verifying theoretical models that enable qualitative and quantitative assessment of climatic change and local forecasting of meteorological parameters in urban agglomerations.

6.3.2 Basic Features of Local Urban Climate

The specific character of a local urban climate can be described in terms of deviations from natural conditions of temperature, humidity, radiation, wind fields and other meteorological parameters. Changes in distribution of these elements are due to the influence of terrain changes (urban buildings), the specific urban atmosphere composition (air pollution), anthropogenic heat sources in the urban areas, etc.

6.3.2.1 Wind Field Structure

Buildings are the basic factors that determine the wind field distribution in urban areas. Air flow over a town may be divided into three areas (Figure 6.3.1). The first one (to about double building-height) usually consists of a closed circulation within the wake of the buildings. The second area, over the first one, extends to the upper limit of an inner boundary layer over the town. The inclination of the inner boundary layer upper limit is of about 0.1 (Panofsky and Townsend, 1964). The third area extends upward to the top of the planetary boundary layer. The wind field in this area is assumed to be that of the upwind rural regions.

Wind distribution in the urban boundary layer depends mainly on the roughness that is mesoscale in character due to the buildings. The mesoscale roughness reduces wind speed at the ground because of obstructions (i.e., buildings) to the air flow. Wind speed reduction is estimated to be 20-30% in comparison with rural regions (Landsberg, 1956). As a result of turbulent viscosity forces the influence of terrain is "diffused" to upper layers, the terrain effect decaying with increasing altitude due to energy dissipation.

The wind speed reduction over a town may be characterized in terms of a power law formulation:

$$u(z) = G\left(\frac{z}{H}\right)^{\alpha}$$

where u = wind speed,

z = height,

G = geostrophic wind speed,

H = geostrophic wind height.

α values estimated according to Sadeh et al., (1973) are as follows:

$$\alpha = \begin{cases} 0.15 \text{ in rural regions,} \\ 0.28 \text{ in suburbs} \\ 0.45 \text{ in town centre.} \end{cases}$$

The results of wind tunnel tests indicate that wind distribution over a town depends on the type and density of buildings and hence, on the direction of advection (Sadeh et al., 1973).

The roughness increase in urban conditions brings about differences in the surface layer wind direction (for a given geostrophic wind direction) in a town relative to its surroundings. The urban area experiences an increase in the cross-isobaric angle in the surface layer. This phenomenon causes a characteristic backing of the wind in many towns in comparison with their environs (Pastorgujeva, 1974).

The dynamic structure of the boundary layers in the urban area, particularly in the surface layer, is undoubtedly influenced by the traffic that brings about wind speed changes and increase in turbulence (Nowak and Sorbjan, 1976).

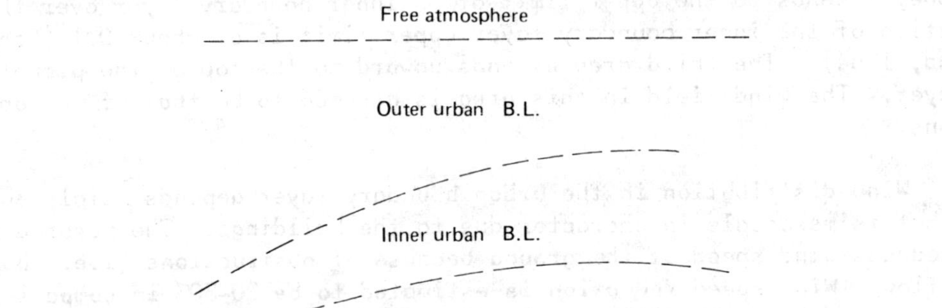

Figure 6.3.1 Urban boundary layer divisions

6.3.2.2 Thermal Structure

Numerous research results indicate that the air temperature within urban areas is on an average a few degrees higher than in the rural environs (Chandler, 1970; Clarke and McElroy, 1970; Landsberg, 1956; Oke, 1974; Oke and Maxwell, 1975; Oke and Hannell, 1970). This phenomenon, usually defined as an urban heat island, is due to artificial heat sources within urban areas, thermal properties of the materials of buildings and streets, evaporation reduction and reduction of heat advection due to obstruction of air flow in the surface layer. Energy balance features of various town surfaces can be found in Oke and Maxwell (1975), Kraujalis-Stoczkowa (1972), and Terjung (1971).

The heat island phenomenum leads to a thermal stratification change in a lower part of the urban boundary layer. When a stable equilibrium exists in the rural regions, neutral or unstable equilibria are usually observed over a town. The phenomenon usually occurs to the height of 300-150 m (Figure 6.3.2). Research results indicate that greatest temperature differences between a town and its vicinity occur 4-5 hours after sunset. The differences begin decreasing thereafter and may even experience a sign reversal. Nocturnal cooling in an urban area is only slightly higher than out of town. Cooler areas are sometimes observable during day hours, too. The occurrence of such coolness enclaves is closely connected with urban air pollution. The air over a town contains many chemically different kinds of dust, nitric oxides, sulfur oxides, carbon monoxide and other products of industrial, communal and traffic origin. The presence of such substances in the air results in attenuating short-wave solar radiation and long-wave radiation of atmospheric and terrestrial origin. Atwater's (1970) evaluation indicates that substances used to describe the radiation field structure of urban areas include water vapor, carbon dioxide and aerosols. The aerosol particle-size distribution in an urban atmosphere is usually described in terms of modified gamma distribution:

$$n(r) = a\, r^{\alpha} \exp\left(-b\, r^{\gamma}\right) \qquad (6.3.2)$$

Atwater (1971) assumes for urban atmospheric aerosols:

a = 5.655×10^{18} (μm^{-4} cm^{-3}),

α = 3.0,

b = 45.1272 ($\mu m^{-1/4}$),

r = particle radius (μm),

$N(r)$ = number density, i.e., number per unit radius interval per unit radius interval per unit volume (μm^{-1} cm^{-3}).

If $N = \int_0^{\infty} n(r)dr$ is the concentration (cm^{-1})

then Atwater's assumptions correspond to N = 10^5 particles per cm^3 and a model radius r_c = 0.005 μm.

The radiation flux in the short-wave range (0.3 - 1 μm) is about 15-20% lower in town that in its vicinity (Landsberg, 1970). Direct radiation decrease and scattered radiation increase are characteristic features of this phenomenon. According to Sprigg and Reifsnyder (1972), direct radiation loss was found to be 20-25%; however, approximately 75-80% of the radiation depleted in the direct beam was regained at the surface as scattered radiation.

Albedo is an important factor in the radiation balance of the earth's surface. The results present in Kung et al., (1964) indicate that urban albedos are about 10% lower than those for rural areas. One factor affecting this difference is the existence of street "canyons" which act as solar radiation traps.

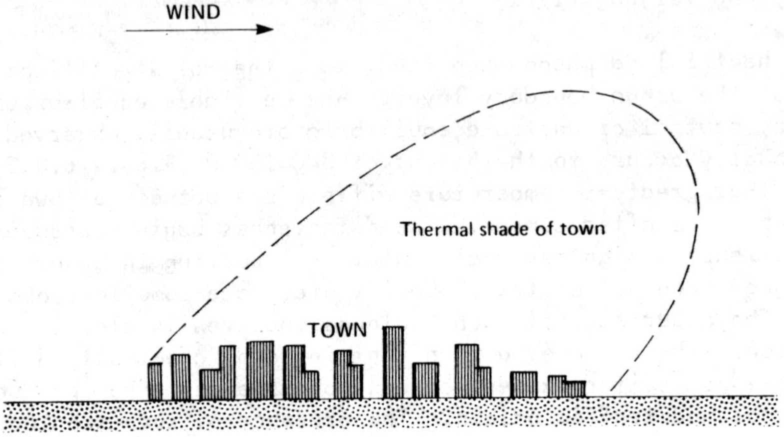

Figure 6.3.2 Thermal structure of the urban BL.

6.3.3 Urban Boundary Layer Modelling

The above-described peculiarities of the urban local climate are explained in many theoretical works that can be divided into two groups, viz., those models which consider separately the thermal and dynamic structures of the urban atmosphere and those which consider the joint thermal-dynamic structure of the air flow over a town.

6.3.3.1 Mathematical Models of the Thermal Structure of the Urban Boundary Layer

Among the thermal structure models of the urban boundary layer we may distinguish models based on the energy balance of the town surface only and models based on the thermodynamic equation. Models described in Leahey (1969), Leahey and Friend (1971), Myrup (1969) and Summers (1965) belong to the former group. They only approximately represent the thermal processes and do not take into consideration the urban atmospheric temperature distribution in space and time (except, in some cases, for specification of the lapse rate).

The energy balance equation for the town surface may be expressed by:

$$(S_b + S_r)(1-a) + F\downarrow - F\uparrow + Q_s = Q_t + Q_e + Q_g \qquad (6.3.3)$$

where S_b, S_r = fluxes of the short-wave direct and scattered solar radiation,

$F\downarrow$, $F\uparrow$ = flux of long-wave radiation from and to the earth surface,

Q_s = anthropogenic heat flux,

Q_t = turbulent heat flux,

Q_e = heat flux due to evaporation,

Q_g = soil heat flux,

a = albedo

Models of the energy-balance type usually solve (6.3.3), or a simplified form thereof, for the urban ground temperature. The solution may involve an iterative procedure.

Models based on the thermodynamic equation

$$\frac{dT}{dt} = \nabla \cdot (K_H \nabla T) + Q_r \qquad (6.3.4)$$

where K_H is the thermal eddy diffusivity coefficient (the vertical part of which is different from the horizontal part),

Q_r is the radional heat flux,

describe the urban atmospheric temperature distribution in space and time. A simplified form of equation (6.3.4) is often used. These models generally require the use of equation (6.3.3) as a lower boundary condition and some additional definition of the dynamic structure of the urban boundary layer. They do not, however, take into consideration the specific character of air flow over the urban area, assuming instead that the wind and turbulence features are the same as over the upwind rural regions.

When the radiational heat influx is taken into consideration, the system of equations (6.3.3) and (6.3.4) needs to be closed by adding the equation of short- and long-wave radiation in the urban atmosphere and a definition of the transmission function due to gas- and dust-atmosphere constituents over a town. The difference in values of Q_R between a town and its surroundings is primarily due to the aerosol layer in the urban atmosphere. The specific character of radiation in the urban area is taken into consideration in Atwater (1970, 1971, 1975); Kagan and Kljagina (1976) and Pandolfo and Jacobs (1973).

6.3.3.2 Mathematical Models of the Dynamic Structure of the Urban Boundary Layer

According to what has been mentioned above, peculiarities of the air flow over a town result from the urban buildings' structure. There are two ways of considering the influence of the buildings on the dynamic structure of the urban boundary layer. The first consists in utilizing an adequately increased roughness parameter representing the dynamic properties of the town surface. The opinion is that the roughness parameter for the urban area should be of order of 1 m. This method is demonstrated by way of example in Atwater (1975); Bornstein (1975) and Pandolfo and Jacobs (1973).

There is, however, a more complete method of describing the dynamic structure of the urban boundary layer. This method can be found in works that take the same approach as towards the description of air flow in vegetation regions (e.g., Menžulin, 1970). It is typical for these models that the urban atmospheric equations embody the components describing the terrain obstructions resistance. The equations of motion can be made to include components describing the resisting force of the buildings:

$$\vec{F}_o = c_r\, |\vec{V}|\, \vec{V}\, I_h \qquad (6.3.5)$$

where c_r = buildings resistance factor,

$\vec{V}$ = wind velocity,

$$I_h = \begin{cases} 1 \text{ for } z \leq h_b, \\ 0 \text{ for } z > h_b, \end{cases}$$

h_b = buildings height,

and, in the case of closing the boundary layer equations by the turbulent energy equation, the latter includes a component characteristic of turbulent energy generation by the resistance of obstructions which is given by:

$$\left(\frac{db}{dt}\right)_r = \vec{F}_o \cdot \vec{V} = c_r\, |\vec{V}|^3\, I_h \qquad (6.3.6)$$

where b = turbulent energy,

t = time

r = (subscript) = building resistance.

This method of describing the dynamics of the urban boundary layer permits a description of the wind distribution in the urban surface layer as well as in the inner and outer boundary layers (Figure 6.3.1).

Assuming stationary, horizontally homogeneous and barotropic conditions the equation of motion of the urban boundary layer takes the form:

$$\vec{T} + \vec{C} + \vec{P} = 0 \qquad (6.3.7)$$

where

$\vec{T} = \frac{\partial}{\partial z}\left(K_M \frac{\partial V}{\partial z}\right) = F_o$ is the friction force,

$\vec{C} = -f\vec{k} \times \vec{V}$ is the Coriolis force,

$\vec{P} = +\, f\vec{k} \times \vec{G}$ is the pressure gradient force,

K_M is the vertical eddy viscosity coefficient,

$\vec{F}_o$ is the resistance force of the buildings (Equation 6.3.5)

f is the Coriolis parameter,

$\vec{k}$ is the unit vector in the vertical,

$\vec{V}$ is the urban wind velocity,

$\vec{G}$ is the geostrophic wind velocity.

Equation (6.3.7) expresses an equilibrium state in which the forces of friction, Coriolis and pressure gradient are in balance. A similar equation could be written for the rural boundary layer. Assuming that the pressure gradient force is the same in both urban and rural environments, then the addition of the buildings' resistance force in the urban area would upset the balance of forces existing outside the town. Consequently, both the Coriolis force and the friction force would have to re-adjust to a new equilibrium state expressed by Equation (6.3.7). This process of adjustment would involve a change to the wind velocity field (otherwise the Coriolis force would remain constant, forcing the friction force to be unchanged also). Examples of this method of the urban boundary layer description can be found in Bykova (1973), Gavrilov (1973) and Sorbjan (1976a).

Sorbjan (1976b) presents a method of considering the influence of traffic on the dynamic structure of the urban boundary layer. The method consists of including within the turbulent energy equation a component analogous to (6.3.6) in the form of:

$$\left(\frac{db}{dt}\right)_s = C\, n\, w^3 \qquad (6.3.8)$$

where C = constant,

n = traffic density,

w = mean vehicle speed,

s (subscript) = traffic

Equation (6.3.8) defines the conversion of kinetic energy of the moving vehicles into atmospheric turbulent kinetic energy.

The wind velocity change would in turn, cause a secondary contribution to change in the friction force due to the presence of the wind velocity in both its terms. The distribution of forces in both rural and urban regions is illustrated schematically in Figure 6.3.3 where, for simplicity, it has been assumed that the urban and rural friction forces are in the same direction. It should be noted that the friction forces are not necessarily in the direction opposite to the respective wind vectors. Figure 6.3.3 shows that the increase of friction force in the town is closely related to the change in direction in which the Coriolis force acts. This change in direction is accompanied by a decrease in magnitude. These changes in the Coriolis force are equivalent to a wind speed decrease and a backing of the wind vector.

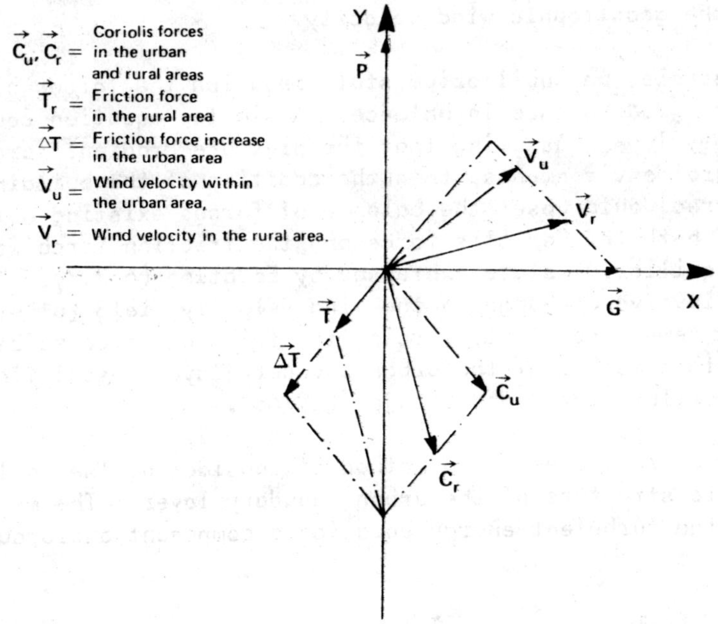

Figure 6.3.3 Force distribution in a town and its vicinity.

$\vec{C}_u$, $\vec{C}_r$ = Coriolis forces in the urban and rural areas

$\vec{T}_r$ = friction force in the rural area

$\vec{\Delta T}$ = friction force increase in the urban area

$\vec{V}_u$ = wind velocity within the urban area,

$\vec{V}_r$ = wind velocity in the rural area.

6.3.3.3 Hydraulic Models of the Urban Boundary Layer

Analysis of data from wind tunnel measurements yields important information about the urban boundary layer structure. Modelling of the air flow over single buildings or areas covered with buildings gives some information about the nature of the wind distribution as a function of space and about the spectrum of turbulence. Diffusion of air pollutants in urban areas is often studied in wind tunnels, too. These pollutants come from surface sources (e.g., traffic) and point sources (e.g., chimneys). These problems are covered in Chaudhry and Cermak (1971); Gorlin and Zpaževskij (1968); Kawatani et al., (1971) and Sadeh et al., (1973).

6.3.4 Conclusions

Great interest in the urban boundary layer structure has been observed recently. Most references cited in this survey date from the last decade. The interest results from rapid growth of urban agglomerations all over the world in recent years. Development of urban areas generates many practical problems of environmental protection. Consequently, numerous research projects of great importance are currently in progress in the field of urban climate measurements and modelling.

Most of the research discussed above have a practical aspect, viz., supplying a large amount of very useful data for mesoscale forecasting of atmospheric air pollution concentration in the urban agglomerations.

Studies are under way to design general-purpose urban atmosphere models that will enable consideration of all important urban effects and definition of temperature, wind, humidity and other meteorological parameters as functions of space and time within the urban areas.

References

Atwater, M.A., 1970: Investigation of the radiation balance for polluted layers of the urban environment. Ph.D. Thesis. New York University, 116 pp.

Atwater, M.A., 1971: Radiative effects of pollutants in the atmospheric boundary layer. *J. Atmos. Sci.*, 28, 1367-1373.

Atwater, M.A., 1975: Thermal changes induced by urbanization and pollutants. *J. Appl. Meteor.*, 14, 1061-1071.

Bornstein, R.D., 1975: The two-dimensional URBMET urban boundary layer model. *J. Appl. Meteor.*, 14, 1459-1477.

Bykova, L.P., 1973: Opyt rasčeta charakteristik pograničnogo sloja atmosfery po zadannym parametram podsloja šerohovatosti. *Tr. G.G.O.*, 297, 12-19.

Chandler, T.J., 1968: Selected bibliography on urban climate. WMO, Publ. No. 276, 383 pp.

Chandler, T.J., 197: Urban climatology - Inventory and prospect. WMO Tech. Note No. 108, 2-14.

Chandler, T.J., 1976: Urban climatology and its relevance to urban design. WMO Tech. Note No. 149, 61 pp.

Chaudhry, F.H. and J.E. Cermak, 1971: Wind-tunnel modelling of flow and diffusion over an urban complex. Tech. Rept. CER70-71 FHC-JEC 24, Fluid Dynamics and Diffusion Laboratory, Colorado State University, Fort Collins, Colorado.

Clarke, T.F. and F.G. McElroy, 1970: Experimental studies of the nocturnal urban boundary layer. WMO Tech. Note No. 108, 108-112.

Gavrilov, A.S., 1973: On the structure of the atmospheric boundary layer above a surface with arbitrary properties of roughness. *Meteorologija i Gidrologija*, 12, 35-42.

Gorlin S.M. and I.M. Zraževskij, 1967: Izučenije obtekanija modelej rel'efa i gorodskoj zastrojki v aerodinamičeskoj trube. *Tr. G.G.O.*, 234, 45-59.

Kagan, R.L. and L.P. Kljagina, 1976: O rasčete potokov korotkovolnovoj radiacji v uslovijach gorodskoj zastrojki. *Tr. G.G.O.*, 365, 61-75.

Kawatani, T., J.E. Cermak and R.N. Meroney, 1971: Characteristic of the mean flow over a simulated urban area. Proc. 3rd International Conf. on Wind Effects on Building and Structures, Toyko, 115-124.

Kraujalis-Skoczkowa, M.W., 1972: Udial ciepla sztucznego w bilansie cieplynm na obszarze Polski. Prz. Geogr. I.G. PAN, 35.

Kung, E.C., R.A. Bryson and D.H. Lenschow, 1964: Study of a continental surface albedo on the basis of flight measurements and structure of the earth's surface cover over North America. *Mon. Wea. Rev.*, *92*, 543-564.

Landsberg, H.E., 1956: The climate of towns. In. Intl. Symp. on Man's Role in Changing the Face of the Earth. Proc. University of Chicago press, 584-606.

Landsberg, H.E., 197: Climate and Urban Planning. WMO Tech. Note 108, 364-374.

Leahey, D.M., 1969: Mathematical models of urban air pollution dynamics; Vol. III, An urban heat island model. New York Univ. School of Engineering and Science. Research Division Geophysical Sciences Laboratory. Report No. TR-69-11, 78 pp.

Leahey, D.m. and J.P. Friend, 1971: A method for predicting the depth of the mixing layer over an urban heat island with application to New York City. *J. Appl. Meteor.*, *10*, 1162-1173.

Menžulin, G.V., 1970: K metodike rasčeta meteorologičeskogo režima v rastitel'nom soobščestvie. *Meteorologija i Gidrologija*, *1970*, *2*, 92-99.

Myrup, L.O., 1969: A numerical model of the urban heat island. *J. Appl. Meteor.*, *8*, 908-918.

Nowak, D. and Z. Sorbjan, 1976: Influence of traffic on dynamical characteristics of the surface layer. Acta Geophisica Polonica, XXIV (1), 57-61.

Oke, T.R., 1974: Review of urban climatology, 1968-1973. WMO Tech. Note 134, 134 pp.

Oke, T.R. and G.B. Maxwell, 1975: Urban heat island dynamics in Montreal and Vancouver. *Atmos. Enviro.*, *9*, 191-200.

Oke, T.R. and F.G. Hannell, 1970: The form of the urban heat island in Hamilton, Canada. WMO Tech. Note 108, 113-126.

Pandolfo, J.P. and C.A. Jacobs, 1973: Tests of an urban meteorological-pollutant model using co validation data in the Los Angeles Metropolitan area. Rpt. No. EPA-R4-73-025a. Met. Lab. Nat. Envir. Res. Cent., North Carolina.

Panofsky, H.A. and A.A. Townsend, 1964: Change of terrain roughness and the wind profile. *Quart. J. R. Met. Soc.*, 90, 147-155.

Pastorgujeva, G.P., 1974: Charakteristika meteorologičeskogo režima prizemnogo sloja vozduha za period ekspedicji v Zaporože. *Tr. G.G.O.*, 332, 17-32.

Sadeh, W.Z., J.E. Cermak and F. Asce, 1973: Turbulence over urban roughness. Pres. at ASCE Nat. Struct. Eng. Meeting, San Francisco.

Sorbjan, Z., 1976a: Numerical model of the structure of urban boundary layer. Ph.D. Thesis. Unpubl. Politechnika Warszawska, Warsaw.

Sorbjan, Z., 1976b: Parameterization of a dynamical structure of boundary layer in urban areas. WMO 450, 27-31.

Sprigg, W.A. and W.E. Reifsnyder, 1972: Solar-Radiation attentuation through the lowest 100 meters of an urban atmosphere. *J. Geophys. Res.*, 77, 6499-6507.

Summers, P.W., 1965: An urban heat-island model - its role in air pollution problems with application to Montreal. Pres. to the First Canadian Conference on Micrometeorology, Toronto.

Terjung, W.H., 1971: The effect of a cyclonic storm on the energy fluxes at the urban interface - a preliminary experiment. *Arch. Met. Geoph. Biokl.*, B, 19, 367-416.

6.4 Electromagnetic and Acoustic Wave Propagation

The propagation of electromagnetic and acoustic waves through the atmosphere is influenced by the turbulence and mean properties of air as a medium. The waves will be refracted and reflected and their energy scattered. The refractive index n for radio frequencies is given by the relationship

$$(n-1)\,10^6 = a_1 p/T - a_2 e/T + a_3 e/T^2 = N \qquad (6.4.1)$$

p being the total atmosphere pressure, e the partial pressure of water vapour, T the air temperature (K), and a_1, a_2, a_3 constants to be determined experimentally. Bean and Dutton (1966) have given values of $a_1 = 77.6$, $a_2 = 5.6$ and $a_3 = 3.75 \times 10^5$. For most atmospheric conditions the equation can be simplified to:

$$N = \frac{77.6}{T}\left(p + \frac{4810e}{T}\right)$$

with a maximum error of 0.1 N unit.

The speed of sound wave propagation with respect to motionless air is under atmospheric conditions given as

$$c = \sqrt{\gamma RT} \qquad (6.4.2)$$

where the specific gas constant R and the ratio of the specific heat capacity at constant pressure to the specific heat capacity at constant volume γ are slightly dependent on the specific humidity of the air. The scattering of sound waves was discussed in Section 5.1.3.

Therefore, the geometry of electromagnetic and acoustic wave propagation in the atmosphere is determined by the peculiarities of temperature and humidity stratification, the latter playing a significant role for radiowaves, particularly in the cm-to-m wavelength range, whereas the sound propagation within the moving air is distinctly influenced additionally by the wind field. It is obvious from this that special phenomena of wave propagation must be expected to occur in the boundary layer of the atmosphere which is characterized by pronounced temperature, humidity and wind profile conditions, including steep gradients near the surface and frequently occurring stable layers (mostly capping inversions) at the top of the boundary layer, which are accompanied by a strong decrease of specific humidity with respect to height and in many cases also by wind shear layers.

Furthermore, the turbulent temperature and humidity fluctuations produce turbulent fluctuations of the refractive index for electromagnetic waves and of the sound speed, respectively, causing a scattering of the waves by turbulent inhomogeneities of the medium which produce characteristic amplitude and phase fluctuations of the received waves. Very extensive literature is available on the investigation, both theoretically and experimentally, of the wave propagation through turbulent media, (e.g., Tatarskii, 1967; Monin and Yaglom, 1975; Gossard and Hooke, 1975; Bean and Dutton , 1966), the collected papers by Brocks (1973) and Zancla (1973) and the proceedings of URSI and IUCRM meetings, the review by Gordon (1972) or the papers in Boundary-Layer Meteorology 4 (1973), for example. Turbulence generally produces fluctuations in the phase and amplitude of electromagnetic and sound waves (scintillation of terrestrial and extra-terrestrial optical and radio sources) as well as refraction phenomena which are significant as disturbing factors for optical and radio astronomy, for communication with terrestrial receivers or with artificial satellites and space probes, for terrestrial and satellite goedesy at optical and radio frequencies etc. On the other hand, the received waves contain valuable information on the characteristics of turbulence and on the stratification pecularities along the wave path providing the possibility of useful methods of remote sensing with the aid of electromagnetic and acoustic waves, which have been used very effectively for boundary layer studies (see Chapter 5).

With respect to the boundary layer structure the following phenomena or wave propagation should be mentioned:

(i) In the optical range the most pronounced effects of terrestrial refraction take place within the surface layer due to extreme temperature (and density) gradients near the surface, especially under extremely unstable (steppe or desert areas under strong isolation) stratification conditions. Anomalous refraction effects near the surface or in higher parts of the boundary layer give rise to well-known mirages. The refraction phenomenon was studied mainly above flat surfaces, i.e. above snow, ice and sea, but also in mountainous regions (e.g. Brocks 1973).

(ii) For radio waves in the cm-to-m wavelength range the above-mentioned effects are complicated by the additional effects of the humidity stratification causing pronounced influences of the surface layer on the wave propagation. For example, duct propagation above the sea surface (Jeske 1975) and reflections on elevated inversions of the refractive index, as described by Fischer (1974), are pecularities of radio wave propagation in the atmospheric surface layer. In this connection, however, the ground-based as well as the capping inversions of temperature at the top of the boundary layer play a significant role for the radiowave propagation in this wavelength region (e.g. Klinker 1955).

(iii) The sound propagation in the boundary layer is determined by the temperature distribution with height and by the wind profile, producing, for example, a concave ray of sound under inversion conditions instead of the convex ray under normal lapse rate conditions (see, for example, the distribution of sound rays for a period of build-up and destruction of a surface inversion, calculated by Franck and Sager, 1974).

Effects of focusing of sound rays at the surface under appropriate temperature and wind profile conditions have been described (e.g. Essenwanger 1966) as well as possible focusing effects of wave structure of low level inversions of the sonic boom overpressures measured at the ground (Herbett et al., 1967).

References

Bean, B.R. and E.J. Dutton, 1966: Radio Meteorology. Nat. Bureau of Standards Monograph., 92, 435 pp.

Brocks, K., 1973: Selected papers on terrestrial refraction, air-sea interaction, and radiometeorology. *Hamburger Geophys. Einselschr.*, 17.

Essenwanger, O., 1966: Zur Berechnung von akustischer Fokussierung in der unteren Troposphäre. *Meteor. Rundsch.*, 19, 130-133.

Fischer, K.E., 1974: Influences of free inversions of the modified refractive index in the troposphere on SHF-over-sea propagation within the horizont. (In German). *Ber. Inst. Radiometeor. Marit. Meteor.*, 27, 129 S.

Franck, H., G. Sager, 1974: Zur Schallwegverteilung in der unteren Grundschicht bei Vorhandensein einer Bodeninversion. *Z. Meteor.*, 24, 130-137.

Gordon, W.E., 1972: Radio and non-ionized media. In: Review of Radio Science, ed. by C.M. Minnis and Y. Bogitch, Internat. Un. Rad. Sci., Brussels, 19-56.

Grossard, E.E. and W.H. Hooke, 1975: Waves in the Atmosphere; atmospheric infrasound and gravity waves, their generation and propagation. Elsevier, Amsterdam, 456 pp.

Herbert, G.A., W.A. Hasse and J.K. Angell, 1969: A preliminary study of atmospheric effects on the sonic boom. *J. Appl. Meteor.*, 8, 618-626.

Jeske, H., 1965: Die Ausbreitung elektromagnetischer Wellen im cm- bis m-Band über dem Meer unter besonderer Berücksichtingung der meteorologischen Bedingungen in der maritimen Grenzschicht. *Hamburger Geophys. Einzelschr.*, 6, 160 S.

Klinker, L., 1958: UKW- Fernempfangsbeobachtungen. Ihre Bedeuntung für Meteorologie und Funktechnik. *Abh. Meteor. Hydrolog. Dienst DDR* 5, 35, 66 S.

Monin, A.S. and A.M. Yaglom, 1975: Statistical fluid mechanics: mechanics of turbulence, Vol. 2, MIT Press, Cambr., Mass., London, 874 pp.

Tatarskii, V.I., 1967: Wave Propagation in a Turbulent Medium. Dover, New York, (R.A. Silverman, translator), 285 pp (also published by McGraw-Hill, New York, 1961).

Zancla, A., 1973: Modern Topics in Microwave Propagation and Air-Sea Interaction (Proc. NATO Adv. Study Inst., June 5-14, 1973, Sorrento, Italy). D. Reidel, Dordrecht, Holland, 364 pp.

6.5 Problems of Mountain Meteorology and Climatology Related to the Boundary Layer

Weather and climate in mountain regions are strongly influenced by distinctive interactions between large-scale atmospheric motions and energy fluxes with the mountain relief. Restrictions of space limit this discussion and for complete reviews interested readers are referred to articles by Steinhauser (1966), by Geiger (1969) and by Flohn (1969). More recent research results are to be found in conference proceedings such as those on Alpine (1967, 1971) and Carpathian (1966, 1971, 1972, 1975) meteorology. Here we undertake only a brief examination of some of the outstanding problems.

Broadly, mountains interact with the atmosphere in two ways. Firstly, they act as physical barriers to airflow. The prevailing airstream is forced to find a way over or around a topographic barrier, or it may be entirely blocked, establishing a pressure gradient such that the flow is locally completely stopped or even reversed. This interaction may be termed the adiabatic, or mechanical, type. A second interaction type involves diabatic forcings. Mountains act as elevated sources, and sinks, of heat and moisture and elevated friction sources. Usually these two forcing types are confounded and it is difficult to establish their relative roles in any particular situation.

Mechanical forcing is perhaps the easier to treat theoretically and there is considerable history of observation of these phenomena. The energetics of the flow and topographic situation determine the partition between horizontal and vertical flow deviation. Horizontal deviations result in "gap winds" where the flow is locally strengthened and canalized. A good example occurs with the northerly Mistral wind of the Rhone Valley between the Alps and the Cevennes. Vertical deviations have been studied in considerable detail by Alaka (1960), Nicholls (1973). Gravity waves are often produced which propagate at least into the stratosphere, and many kilometers downstream. Such waves are frequently accompanied by rotor clouds and violent turbulence. Of more importance for the boundary layer is that these orographically induced gravity waves transfer considerable momentum downward in the boundary layer, the flux being probably comparable with the eddy-momentum flux. Thus there is an enhanced drag at the surface in mountain regions. The geostrophic drag coefficient, as predicted by the resistance law of semi-empirical boundary layer theory (see Section 2.3), must be replaced by an increased value over mountain areas. Further studies are needed to quantify this effect.

The diabatic influence on the mountain boundary layer is perhaps most noticeable in the occurrence of pronounced mesoscale wind systems with a marked diurnal variation, see Burman (1969). Thus "slope winds", which blow normal to local topographic contours, are a direct result of the density differential between air on a slope in contract with the surface and that in the ambient atmosphere at the same level. Typically up-slope winds begin shortly after sunrise and give way to down-slope winds around sunset, with wind speed dependent on slope and surface heating. Valley winds blow along the main axis of a valley which opens into an adjacent plain. They are caused by pressure differences due to differential heating of the air in these two regions, see Cross (1950). The complexities of these systems and their interactions are not well understood, but their importance is evidenced by increasing study because of effects on air pollution dispersion.

Such slope and valley winds are manifestations of a large class of phenomena of mountainous terrain caused by the inhomogeneous nature of the underlying surface. In these conditions horizontal advections of heat, moisture and momentum become as important, and in many cases more important, than vertical transports. Some of the resulting pecularities are expressed even in climatological mean values, for example Steinhauser (1966). One illustration is an observation by Whiteman and McKee (1977) of diurnal heating in a mountain valley near Vail, Colorado, resulting in erosion of the nocturnal inversion layer from the top down, rather than the bottom up as is commonly observed over flat terrain. Observations such as this indicate that the important physical mechanisms in mountainous terrains can be quite different from those over homogeneous terrain, and strike at the roots of the usual boundary layer parameterizations applicable to barotropic, stationary and horizontally homogeneous cases.

Apparently such effects can extend horizontally as much as one hundred times the height of the mountains since it is found that the boundary layer vector-wind profile is determined largely by the direction of flow with respect to the mountains. Investigation and understanding of such inhomogeneous conditions remains one of the most challenging areas of the science.

Characteristic meteorological conditions of mountainous regions have important consequences for the ecology of these regions. Distinctive variations, such as those of vegetation with slope, aspect and altitude are well known, see for example Schneider-Carius (1953) and Weischet (1965, 1966). Mention has already been made of important constraints on turbulent dispersion, which have become an increasing concern with industrialization. It is evident that mountains give rise to unique atmospheric boundary layer features. The applicability of knowledge gained for flat-land situations is severely limited. Much new study is required to advance knowledge of this important environment.

References

Burman, E.A., 1969: Local winds. (In Russian). Hydromet. Press, Leningrad, 341 pp.

Cross, C.M., 1950: Slpe and Valley Winds in the Columbia River Valley. *Bul. Amer. Meteor. Soc.*, 31, (3): 79-74.

Flohn, H., 1969: Local Wind Systems; World Survey of Climatology. Volume 2. Elsevier Publishing Co., New York, 266 pp.

Geiger, R., 1969: Topoclimates; World Survey of Climatology. Volume 2. Elsevier Publishing Co., New York, 266 pp.

Nicholls, J.M., 1973: The airflow over mountains - Research 1958-1972. WMO Techn. Note 127 WMO - No. 355, 64 pp.

Schneider-Carius, K., 1953: Die Grundschicht der Troposphäre. Leipzig, 168 S.

Steinhauser, F., 1966: Die planetarische Grenzschicht der Atmosphäre mit besonderer Berücksichtigung der Gebirgsmeteorologie. Meteorologie. Ergenbnisse der Konferenze über Meteorologie in Liblice bei Prag 13.bis 16. Oktober 1964. Academia Press, Prague, 85-111.

Weischet, W., 1965: Der tropisch-konvektive und der außertropisch-advektive Typ der vertikalen Niederschlagsverteilung. *Erdkunde*, 19, 6-14.

Weischet, W., 1966: Zur Klimatologie der nordchilenischen Wüste. *Meteor. Rundsch.*, 19, 1-7.

Whiteman, C.D. and T.B. McKee, 1977: Observations of Vertical Atmospheric Structure in a Deep Mountain Valley. *Arch. Met. Geoph. Biokl., Ser. A*, 26, 39-50.

1967:9. Internationale Tagung für Alpine Meteorologie, Brigg und Zermatt 14. - 17. September 1966. Veröff. Schweizer. *Meteor. Zentralanstalt*, *4*, 366 S.

1971:XI. Internationale Tagung für Alpine Meteorologie in Oberstdorf 22.-26. September 1970. *Ann. Meteor., Neue Folge*, 5, 300 S.

1966:Einfluß der Karpaten auf die Witterungserscheinungen. III. Konferenze für Karpatenmeteorologie Beograd 27.-30. Mai 1965, Beograd, 418S.

1971:IV. Miezdynarodowa Konferencja Meteorologii Karpackiej, Krakow 16.20. Wrzensnia 1969. *Prace Geograficzne*, 26, Krakow, 428 pp.

1972:V. Konferenze für Karpatenmeteorologie, 14.-20. IX 1971, Bucuresti, 539 S.

1975:Die Vorträge der VI. Internationalen Konferenzuber Karpatenmeteorologie 17.-22. September 1973. Kiev, 615 S.

APPENDIX A

REFERENCES - (See Chapters 2 and 3 for additional references)

Agee, E.M., D.E. Brown, T.S. Chen and K.E. Dowell, 1973: A height-dependent model of eddy viscosity in the planetary boundary layer. *J. Appl. Meteor.*, 12, 409-412.

Beryland, M.E., 1947: The theory of the wind variation with respect to height. (In Russian). *Trudy NIU GUGMS, Ser. 1*, 25, 14-67.

Blackadar, A.K., 1965: A simplified two-layer model of the baroclinic neutral atmospheric boundary layer. Final Rep. AFCRL 65-531, Pennsylvania State Univ., 49-65.

Blinova, E.N. and I.A. Kobel, 1937: Application of the theory of turbulence to the question of distribution of wind with respect to height. (In Russian). In: Izvekov, B.I., Kochin, N.E.: Dynamic Meteorology, Part 2, Hydromet. Press, Leningrad, Moscow, 28-36.

Ellison, T.H., 1956: Atmospheric turbulence. In: Surveys in Mechanics, Ed. by G.K. Batchelor and R.M. Davies, Cambr. Univ. Press, 400-430.

Frost, R., 1948: Atmospheric turbulence. *Quart. J. Roy. Meteor. Soc.*, 74, 316-338.

Godev, N. and D. Yordanov, 1967: On models for the study of wind variation with height in the planetary boundary layer. (In Bulg.). *Diklady Bolgar. Acad. Sci.*, 20, No. 9, 911-913.

Klyuchnikova, L.A., D.L. Laikhtman and G. Kh. Tseitlin, 1965: On the question of the calculation of the vertical wind profile in the atmospheric boundary layer. (In Russian). *Trudy Main Geophys. Obs. (GGO)*, 167, 3-28.

Köhler, H., 1933: Meteorologische Turbulenzuntersuchungen. Kungl. Sve. Wetenkapsakad. Handl. Tredje Ser. 13, 1-54.

Kovetz, A., J. Neumann and R. Suzi, 1963: Extension of Ekman's spiral to conditions of near-neutral static stability and a constant horizontal temperature gradient. *Beitr. Phys. Atm.* 36, 201-212.

Prandtl, L. and W. Tollmien, 1924/25: Die Windverteilung über dem Erdboden, errechnet aus den Gesetzen der Rohrströmung. *Z. Geophys.* 1, 47-55.

Sakakibara, S., 1982: On the traverse eddy resistance acting on moving air in the lower atmosphere. *Geophys. Mag.* 1, 4, 130-149.

APPENDIX A

REFERENCES - (Continued)

Shvets, M.E., 1941: The determination of the coefficient of eddy viscosity for atmospheric motions. (In Russian). *Doklady Acad. Sci. USSR*, 30, 8, 705-709.

Shvets, M.E. and M.I. Yudin, 1940: A stationary model of wind distribution with respect to height in a turbulent atmosphere. (In Russian). *Trudy Main Geophys. Obs. (GGO)*, 31, 42-52.

Takaya, S., 1930: On the coefficient of eddy viscosity in the lower atmosphere. *Mem. Marine Obs. Kobe, Japan 4*, 1, 1-33.

Takev, K., 1964: The wind profile for a power model of the law of kinematic eddy viscosity coefficient with height. (In Bulg.). *Khidrologiya i Meteorologiya*, 3, 3-9.

Tseitin, G. Kh. and L.R. Orlenko, 1960: The stationary distribution of wind temperature, and turbulent exchange in the boundary layer under various stability conditions. (In Russian). *Trudy Main Geophys. Obs. (GGO)*, 94, 8-28.

APPENDIX B

REFERENCES - (SEE CHAPTER 2 FOR ADDITIONAL REFERENCES)

Arkhangelskii, V.L., 1955: On the friction height under moderate mountainous relief conditions. (In Russian). *Meteorologiya i Gidrologiya* 5, 26-28.

Baumgartner, A., 1956: Untersuchungen über den Wärme- und Wasser-haushalt eines jungen Waldes. *Ber.-Deu. Wetterd.,* 5, 28, 3-53.

Bernstein, A.B., 1959: The effect of a horizontal temperature gradient on the surface wind. M.S. Dissertation, Pennsylvania State University.

Brown, R.A., 1974b: A simple momentum integral model. *J. Geophys. Res.,* 79, 4076-4079.

Frankenberger, E.F.H., 1952: Investigations of the vertical exchange within the lowest decameters at Quickborn. *Geophys. Res. Pap. No.* 19, GRD, Air Force Cambr. Res. Center, 323-340.

Johnson, W.B., 1965: Atmospheric boundary-layer dynamics over the forests of northeastern Wisconsin. Ph.D. Thesis, Dept. Meteor., Univ. of Wisconsin.

Lettau, H., 1950: A re-examination of the "Leipzig wind profile" considering some relations between wind and turbulence in the frictional layer. *Tellus* 2, 125-129.

Lettau, H., 1958: Windprofil, innere Reibung und Energieumsatz in den unteren 500 m über dem Meer. *Beitr. Phys. Atm.,* 30, 78-96.

Lettau, H., 1959: Wind profile, surface stress and geostrophic drag coefficients in the atmospheric surface layer. *Adv. Geophys.* 6, 241-256.

Lettau, H., H. Howeber, Über die Bestimmung der Höhenverteilung von Schubspannung und Austauschkoeffizient in der atmosphärischen Reibungsschicht. *Beitr. Phys. Atm.,* 37, 105-118.

Sawyer, J.S., 1959: The introduction of the effects of topography into methods of numerical forecasting. *Quart. J. Roy. Meteor. Soc.,* 85, 31-43.

Seeliger, W., 1938: Höhenwind und Gradientwind. *Beitr. Phys. fr. Atm.,* 24, 130-148.

Sutcliffe, R.C., 1936: Surface resistance in atmospheric flow. *Quart. J. Roy. Meteor. Soc ,* 62, 3 12

Sverdrup H.V., 1957: Norsk. Polarinst. Skrifter, Oslo 111.

APPENDIX I

SUMMARY OF K-MODELS

Author	K	ℓ	Range of Validity	Constants
One-layer models				
Ekman (1905, 1906) Akerblom (1908), Taylor (1915) Exner (1912), Sakakibara (1928) MacKay (1971)	K_M = constant	-	-	-
Prandtl and Tollmien (1924-25) Kohler (1933), Frost (1948) Takev (1964)	$K_M \sim z^P$	-	-	-
Moller (1931), Blinova and Kibel (1937) Ellison (1956)	$K_M \sim z$	-	-	-
Takaya (1930)	$K_M \sim z^2$	-	-	-
Kovetz, Neumann and Suzi (1963)	$K_M = \frac{ku_* z}{1 + Bz + aB^2 z^2}$	-	-	$B = \beta/L \quad \beta \sim 6$
Rossby (1932)	$K_M \sim (1 - z/z_*)^2$	-	$z \leq z_*$	-
Takaya (1930)	$K_M \sim e^{az}$	-	-	-
Shvets (1941)	$K_M \sim 1 - e^{-az}$	-	-	-
Agee et al. (1973)	$K_M \sim e^{-bz/z_T} - e^{-bcz/z_T}$	-	-	-

APPENDIX I - (CONTINUED)

Author	K	ℓ	Range of Validity	Constants
Misra (1977)	$K_M = K_o\,(1 - z/h)^p$	-	-	$1 \leq p \leq 2$
Busch & Chang (1976)	$K_M = A^{-1/2}\, b^3 \ell^2 \left(S^2 - \frac{g}{\Theta}\frac{K_H}{K_M}\frac{\partial\Theta}{\partial z}\right)^{1/2}$ $K_H/K_M = \phi_M/\phi_H$ (Based on stationary TE-equation)	$\frac{\partial \ell}{\partial t} = \frac{\ell_s - \ell}{\ell/(u_*^2 + w_*^2)^{1/2}}$ $w_* = \left(\frac{g}{\Theta}\,\overline{w\,\Theta_o}\cdot zi\right)^{1/3}$ $\ell_s = \frac{kz}{\phi_M}\left(1 - \frac{z}{z_i}\right)\; z < h'$ $\ell_s = \ell_B \; z > h'$	All Ri	z_i = height of PBL h' = the height where ℓ becomes equal to a residual mixing length ℓ_B
Estoque & Bhumralkar (1970)	$K_M = \ell^2 S^* (1 + \alpha Ri)$ $K_M = \ell^2 S \left(1 - \alpha Ri\right)^{-1}$	$\ell = \frac{kz}{1 + \frac{kz}{\lambda}}$	Ri > 0 Ri $\leq$ 0	$\lambda = 2.7\cdot10^{-4} \left\lvert\frac{V_g}{f}\right\rvert$ $\alpha = -3.0$
Clarke & Clayton (Clarke, 1974)	$K = \ell^2 S\,(1 - \gamma\sigma)$ $K = \ell^2 S\,(1 + \sigma)^{-1}$ $\sigma = \frac{(g\ell)^{1/2}}{\Theta}\frac{\partial\Theta}{\partial z}/S$	$\ell = \frac{kz}{1 + \frac{kz}{\lambda}}$	$\partial\Theta/\partial z < 0$ $\partial\Theta/\partial z > 0$	$\gamma = 18$ $\lambda = 2.7\cdot10^{-4}\cdot\left\lvert\frac{V_g}{f}\right\rvert$

$$^*S^2 = \left(\frac{\partial u}{\partial z}\right)^2 + \left(\frac{\partial v}{\partial z}\right)^2, \quad Ri = \frac{\frac{g}{\Theta}\frac{\partial\Theta}{\partial z}}{S^2}$$

APPENDIX I - (CONTINUED)

Author	K	ℓ	Range of Validity	Constants
Geleyn (1975)	$K_M = \ell^2 \left(S^2 - \frac{g}{\Theta}\frac{\partial\Theta}{\partial z}\right)^{1/2}$ $K_M = 0$ when this factor ≤ 0	$\ell = \left(kK_M/\ell^2\right)/\frac{\partial}{\partial z}\left(K_M/\ell^2\right)$	All Ri	-
Karlsson (1972)	$K_M = \ell^2 S\,(1 + \alpha Ri)^{-2}$ $K_H = K_M/(1 + \alpha Ri)$	$\ell = \frac{kz}{1 + \frac{kz}{\lambda}}$	$\frac{\partial\Theta}{\partial z} \geq 0$	$\alpha = 3.0$ λ = same as Clarke & Clayton
Shir (1973)	$K_M = ku_* z \cdot \frac{1}{2}\left[e^{-4z/h} + \frac{1}{1 + a(z/h)^b}\right]$	-	$0 < z < h$ $Ri = 0$	$a = 16$, $b = 1.6$ $h = 0.455\ u_*/f$
Sorbjan (1976)	$K_M = \{\ell^6 \left(S^2 - \frac{g}{\Theta}\frac{\partial\Theta}{\partial z} I_\Theta\right)^{3/2} + \ell^4 E\}^{1/3}$ (Based on stationary TE-equation)	$\ell = 2.7 \cdot 10^{-4} \left\lvert V_g \right\rvert \cdot s/\lvert f \rvert$	$\partial\Theta/\partial z > 0$ $\frac{\partial\Theta}{\partial z} < 0$	$s = \left(100 \frac{\partial\Theta}{\partial z} + 1.0\right)^{-1}$, $I_\Theta = 0$ $s = 250 \left(\frac{\partial\Theta}{\partial z}\right)^{1/4} + 1.0$ $I_\Theta = 1$ $E = C_\tau \lvert \overline{V} \rvert^3$ addition of TE buildings $E = 0$ above typical building height
Walmsley (1973)	$K_M = \frac{ku_* z}{\phi_M} e^{-z/\beta}$	-	$z_o \leq z \leq z_i$ All Ri	$\beta = z_i/n \quad n \sim 4$ ϕ_M - Businger et al. (1971)

APPENDIX I - (CONTINUED)

Author	K	ℓ	Range of Validity	Constants
Wu (1965)	$K_M = \left(S^2 - \alpha_H \frac{g}{\Theta}\frac{\partial\Theta}{\partial z}\right)^{1/2} \ell^2$ $K_H = \alpha_H K_m$	$\ell = \frac{kz}{1 \quad kz}$	$\lambda = 100$ m $\frac{\partial\Theta}{\partial z} < 0$ $\lambda = 30$ m $\frac{\partial\Theta}{\partial z} \approx 0$ $\lambda = 12.5$ m $\frac{\partial\Theta}{\partial z} > 10^{-4}$	$\alpha_H \sim 1$
Yamamoto & Shimanuki (1966) (Modified in Yamamoto et al., (1973)	$K_M = k^2 z^2 \left(S + \sqrt{\frac{g}{\Theta}\left\lvert\frac{\partial\Theta}{\partial z}\right\rvert}\right)$ $K_M = k^2 z^2 \left(S - \left\lvert\frac{L}{z}\right\rvert^{1/6} \sqrt{\frac{\sigma g}{\Theta}\frac{\partial\Theta}{\partial z}}\right)$	-	$\frac{\partial\Theta}{\partial z} < 0$ $\frac{\partial\Theta}{\partial z} > 0$	$\sigma = 15$
Estoque (1963)	$K_M = \{kz\,(1 + \alpha Ri)\}^2 S$ $K_M = \lambda\left(\frac{g}{\Theta}\right)^{1/2} \left\lvert\frac{\partial\Theta}{\partial z}\right\rvert^{1/2} z^2$ $K_M = K_s\,(z_i - z)/(z_i - z_s)$	- - -	$Ri > Ri_c,\ z \le z_s$ $Ri < Ri_c,\ z \le z_s$ All $Ri,\ z_s < z < z_i$	$Ri_c = -0.03$ $\lambda = 0.9;\ \alpha = -3.0$
O'Brien (1970)	$K_M = K_i + \left(\frac{z - z_i}{z_i - z_s}\right)^2 \{K_s - K_i$ $+ (z - z_s)\left(\left.\frac{\partial K}{\partial z}\right\rvert_s + \frac{2(K_s - K_i)}{z_i - z_s}\right)\}$	-	$z_s < z < z_i$ All Ri^*	z_i top of Bl z_s top of surface layer

APPENDIX I - (CONTINUED)

Author	K	ℓ	Range of Validity	Constants
Two-layer models				
Rossby and Montgomery (1935)	$K_M \sim z$	-	$z_o \leq z \leq z_1$	-
	$K_M \sim \left(1 - \frac{z}{z_*}\right)^2$	-	$z_1 \leq z \leq z_*$	
Shvets and Yudin (1940) Blackadar (1965)	$K_M \sim z$	-	$z \leq z_1$	-
	K_M = constant		$z_1 \leq z$	
Berlyand (1947) Tseitin and Orlenko (1960)	$K_M \sim z^p$	-	$z \leq z_1$	-
	K_M = constant		$z_1 \leq z$	
Klynchnikova, Laik Klynchnikova, Laikhtman and Tseitin (1965)	$K_M \sim z^p$	-	$z \leq z_1$	-
	$K_M \sim e^{-az}$		$z_1 \leq z$	
Godev and Yordanov (1967)	$K_M \sim z^p$	-	$z \leq z_1$	-
	$\sim z^q$		$z_1 \leq z$	
	$K_M \sim z^p$		$z \leq z_1$	
	$\sim e^{az}$		$z_1 \leq z$	

APPENDIX II - GEOSTROPHIC DRAG COEFFICIENT, CALCULATED FROM OBSERVATIONS

Site	Observ. Period	Stratification	V_g ms^{-1}	10^2 C_g	Calculated By	Data Source
Leipzig, GDR	20.10.1931	Nearly Neutral	17.5	3.7	Lettau (1950)	Mildner (1932)
			15.5	4.4	Venkatesh and Csanady (1974)	Mildner (1932)
Middle Europe	July 1935- January 1936	Nearly Neutral	14.8	5.3	Sawyer (1959)	Seeliger (1938)
			14.8	5.8	Rüffer (1976)	Seeliger (1938)
Brookhaven, N.Y.	July 1950	Nearly Neutral	10.0	5.0	Blackadar (1962)	Bernstein (1959)
Lakewood, Wisc.	-	Nearly Neutral	14.1	5.2	Johnson (1965)	
Quickborn, FRG	1952	Nearly Neutral	10.0	3.9	Lettau (1959	Frankenberger (1952)
Hay, Australia	15.7.-28.8. 1967	Nearly Neutral	13.9	3.0	Clarke et al. (1971)	
	15.7.-27.8 1967	Nearly Neutral	5-10 (7.2)	3.2.	Venkatesh and Csanady (1974)	
	15.7.-27.8 1967	Nearly Neutral	$\geq$ 10 (11.1)	2.6	Venkatesh and Csanady (1974)	
Near Munich, FRG	28.6.-5.7. 1952	Nearly Neutral	15.0	4.0	Lettau (1959)	Baumgartner (1956)
Lindenberg, GDR	1962-1964	Nearly Neutral	10.0	6.7	Bernhardt (1974b) (1975)	
	1962-1964	Nearly Neutral	15.0	5.0	Bernhardt (1974b) (1975)	
Canitz near Riesa, GDR	13.9.1960	Nearly Neutral	6.1	6.8	Rüffer (1976)	
O.Neill, Nebraska	8.8.-8.9. 1953	Nearly Neutral	14.5	3.3	Lettau (1959)	
	13.-14.8 1953	Nearly Neutral	10.2	2.2	Venkatesh and Csanady (1974)	
	18.-19.8 1953	Nearly Neutral	9.7	5.0	Venkatesh and Csanady (1974)	
Johnson, Isl.	-	Nearly Neutral	11.6	2.3.	Lettau (1959)	
Cambridge	-	Nearly Neutral	9.0	2.8	Lettau (1959)	

APPENDIX II - (CONTINUED)

Site	Observ. Period	Stratification	V_g ms^{-1}	10^2 C_g	Calculated By	Data Source
North Atlantic	-	Nearly Neutral	11.0	2.7	Deacon (1973)	
Arctic Ice (Aidjex-Experiment)	1972	Nearly neutral	5.2	3.8	Brown (1974b)	
	1972	Neutral to unstable	5.3	4.6	Brown (1974b)	
	1972	Stable	5.4.	2.5	Brown (1974b)	
Western Baltic (Kiel Bight)	October 1967	Neutral to unstable	14.3	2.2	Hasse and Cunckel (1974)	
Scilly Island.	4.1.1951	Unstable	12.2	2.2.	Lettau (1958)	
	10.1.1951	Stable	6.6	2.3.	Lettau (1958)	
Helgoland I	19.8.1960	Unstable	11.0	2.0	Lettau and Hoeber (1964)	
Helgoland II	31.8.1960	Stable	12.7	2.2.	Lettau and Hoeber (1964)	
Hay, Australia	15.7.-27.8 1967	Stable	5-10 (7.7)	2.7	Venkatesh and Csanady (1974)	
	15.7.-27.8 1967	Stable	$\leq$ 10 (12, 6)	2.5	Venkatesh and Csanady (1974)	
Lindenberg, GDR	1962-1964	Stable (inversion grad. surface/850 mbar)	10.0	6.2.	Bernhardt (1974b) (1975)	
			15.0	4.8	Bernhardt (1974b) (1975)	
Meppen, FRG	1971	-	7.2	4.6	Leykauf (1974)	
			14.1	4.1	Leykauf (1974)	
England	-	-	6.0	7.8	Sutcliffe (1936)	
North Atlantic	1925-1927	-	7.8	2.0	Sutcliffe (1936)	
East Siberia, USSR	-	-	-	6.5	Sawyer (1959)	Arkhangelskii (1955)
Central USA	Summer conditions	-	-	6.8	Sawyer (1959)	Buajitti and Blackadar (1957)
Arctic ice	-	-	-	4.0	Sawyer (1959)	Sverdroup (1957)

APPENDIX III - SUMMARY OF MIXING-LENGTH FORMULATIONS

Yamada & Mellor (1975) - {Also used by Yu (1976)}

$$\ell = \frac{kz}{1 + \frac{kz}{\ell_o}}, \quad \ell_o = \alpha \frac{\int_0^\infty qzdz}{\int_0^\infty qdz} \quad \text{where} \quad q = (26)^{1/2}, \; \alpha = 0.1 \tag{1}$$

For both stable $(\overline{w\theta_o} < 0)$ and unstable stratification $(\overline{w\theta_o}) > 0$

Djolov (1973) used in Yu (1976)

$$\frac{1}{\ell} = \frac{1}{\lambda} + \frac{\phi_M(z/L)}{kz} \; ; \quad \overline{w\theta_o} \gtrless 0 \tag{2}$$

$$\lambda = 2.7 \cdot 10^{-4} \cdot |\underline{V}_g| / |f| \tag{3}$$

Delgage (1974)

$$\frac{1}{\ell} = \frac{1}{kz} + \frac{1}{\lambda} + \frac{\beta}{kL} \qquad \overline{w\theta_o} < 0 \tag{4}$$

$$\lambda = 4 \cdot 10^{-4} |\underline{V}_g| / |f| \qquad \beta = 5 \tag{5}$$

Bodin (1976)

$$\frac{1}{\ell} = \frac{1}{\ell_N} - \frac{1}{z_i} \cdot \frac{1 + 1.32 \frac{z}{z_i}}{0.001 + \frac{kz}{z_i}} \cdot \frac{1}{1-50 \frac{L_o}{z_i}} \; ; \quad \overline{w\theta_o} > 0 \tag{6}$$

L_o is the surface Monin-Obukhov length and

$$\frac{1}{\ell_N} = \frac{1}{kz} + \frac{1}{0.2\, z_i} \tag{7}$$

z_i is diagnosed from the heat flux profile

APPENDIX III - (CONTINUED)

$\ell = \ell_N \qquad \overline{w\theta}_o < 0$

with $z_i = 0.06 \cdot u_* / |f|$ (8)

Equation (6) is an interpolation formula suggested by Deardorff.

Busch & Chang (1976)

$$\frac{\partial \ell}{\partial t} = \frac{\ell_s - \ell}{\ell / (u_*^2 + w_*^2)^{1/2}} \tag{9}$$

$$\ell_s = \frac{kz}{\phi_M}\left(1 - \frac{z}{z_i}\right) \quad \text{for } z < h'$$

$$\ell_s = \ell_B \quad \text{for } z > h'$$

where h' is the height where ℓ becomes equal to a residual mixing length, ℓ_B, above the boundary layer.

References: See Chapter 4.